GEOTECHNICAL SPECIAL PUBLICATION NO. 95

SOIL–CEMENT AND OTHER CONSTRUCTION PRACTICES IN GEOTECHNICAL ENGINEERING

PROCEEDINGS OF SESSIONS OF GEO-DENVER 2000

SESSION TRACK CO-SPONSORED BY
The Geo-Institute, Construction Division, and Materials Division
of the American Society of Civil Engineers

August 5-8, 2000
Denver, Colorado

W0017600

EDITED BY
Cliff Schexnayder

GEO
INSTITUTE

ASCE **American Society of Civil Engineers**
1801 ALEXANDER BELL DRIVE
RESTON, VIRGINIA 20191–4400

Abstract: The Construction track proceeding of the August 2000, Geo-Denver, Geo-Institute, American Society of Civil Engineers, Specialty Conference, addresses soil–cement construction procedures and other innovative solutions to geotechnical construction issues. Overall, these proceedings represent a significant contribution to the state of the art for geotechnical construction practice. It will effectively assist practicing engineers, constructors, inspectors, and researchers. Besides several overviews of soil–cement construction practice, there are case studies addressing pile construction, use of post-tension anchors for seismic upgrades, claims issues on a difficult muck excavation project, the relocation of a lighthouse, use of hollow piers for excavation support, and deep excavations in soft soil.

Library of Congress Cataloging-in-Publication Data

Geo-Denver 2000 (2000 : Denver, Colo.)
    Soil-cement and other construction practices in geotechnical engineering : proceedings of sessions of Geo-Denver 2000 : August 5-8, 2000 Denver, Colorado / sponsored by the Geo-Institute of the American Society of Civil Engineers ; edited by Cliff Schexnayder.
        p. cm. – (Geotechnical special publication ; no. 95)
    Includes bibliographical references and index.
    ISBN 0-7844-0500-X
    1. Engineering geology—Congresses. 2. Soil cement construction—Congresses.
3. Soil-cement—Congresses. I. Schexnayder, Cliff J. II. American Society of Civil Engineers. Geo-Institute. III. Title. IV. Series.

TA703.5. G417 2000
624.1'5—dc21

                                                                        00-033140

Geotechnical Special Publications

Contents

Preface

These proceedings include the technical paper presented in the construction and materials tract of the Geo-Denver Specialty Conference in Denver, Colorado, August 5-8, 2000. This conference was sponsored by the Geo-Institute of the American Society of Civil Engineers. The construction and materials tract was co-sponsored and organized by the Construction and Materials Division of ASCE and the Geo-Institute.

This joint sponsorship of sessions addressing issues of concern to these three areas of practice, construction, materials, and geotechnical, began with a roller-compacted concrete symposium at the May 1985 convention of the Society in Denver, Colorado. Therefore, it is very fitting that we return to Denver to address new issues of mutual concern.

The objective of the conference was to foster an exchange of data on innovative practice based on data from actual projects. Presented in these proceedings are case studies addressing soil cement construction practice, pile construction, use of post-tension anchors for seismic upgrades, claims issues on a difficult muck excavation project, the relocation of a lighthouse, use of hollow piers for excavation support, and deep excavations in soft soil. The speakers for the conference were chosen for their experience and expertise in the topics covered.

Together Kenneth D. Hansen and the editor of the proceedings planned this conference tract. The technical papers contained in these proceedings are as furnished by the authors. The ideas are those of the authors and do not necessarily represent the views of the American Society of Civil Engineers. The Proceedings Editor, with the concurrence of one additional outside reviewer, made the decision of accepting papers for publication. All papers are eligible for discussion in the Journal of Construction Engineering and Management, and for ASCE awards.

Cliff Schexnayder

Construction of Stair-Stepped Soil-Cement Bank Protection

By Kenneth D. Hansen[1] and Cliff Schexnayder,[2] Fellows ASCE

Abstract: Along normally dry rivers in the arid southwestern United States soil-cement has become a popular method of protecting banks and bridge abutments from erosion and collapse during floods. This paper focuses on construction of soil-cement bank protection. Construction topics discussed include control of water, central mixing plants, and the equipment needed to transport, spread, compact and cure soil-cement built in stair-stepped fashion on relatively steep riverbanks. Additionally the materials, mixture proportions and section dimensions of soil-cement bank protection, are discussed. Production rates and cost factors are also presented.

INTRODUCTION

The erosion of sand stream banks is a severe problem when there are flood flows in rivers passing through urban areas such as Tucson, Phoenix, and Albuquerque in the arid southwestern United States. During the October 1983 flood at Tucson, unstabilized banks on the Santa Cruz river eroded as far back as 469 ft (140 m) from their pre-flood locations. Figure 1 shows the collapse of the end span of the northbound lane of the Interstate 19 bridge, which is located just south of Tucson. In the background it is seen that the entire bridge on the road leading to the San Xavier Indian Reservation has been washed away. The left sand abutment eroded back several hundred feet to the position shown in the photograph. The abutments at both ends of the Interstate Highway Bridge are now protected with soil-cement bank protection.

The prime method for addressing this problem is the use of cement-stabilized on-site sand, known as soil-cement. This cement-stabilized natural material protects banks and is used to protect grade control structures in these areas.

---

[1] Senior Vice President, Schnabel Engineering Associates, 1888 Sherman St. #330 Denver, CO 80203; phone: (303) 863-0422, E-mail: khansen@schnabel-eng.com
[2] Eminent Scholar, Del E. Webb School of Constr., Box 870204, Arizona State Univ., Box 870204 Tempe, AZ 85287-0204; E-mail: Cliff.s@asu.edu.

The design and construction of soil-cement bank protection is similar to soil-cement slope protection for new earth dams and other embankments. There are differences however, particularly the issue of water control and the steep slope of the embankment to be protected. Both designs involve placing horizontal layers of soil-cement in a stair-step fashion up an embankment slope. Bank protection slopes are typically steeper than those of earth dams. Current designs for the upstream slopes of new embankment dams tend to be about 3.0H:IV. Slopes of 1.0H to 1.0V for soil-cement bank protection conform to the near vertical slopes of the native banks. Such steep slopes minimize right-of-way requirements and increase the hydraulic efficiency of the cross section of the channel.

FIG. 1. Eroded sand banks and damaged bridges on the Santa Cruz River south of Tucson, Arizona.

CONTROL OF WATER

It is very seldom that ground or flowing water is encountered in the construction of soil-cement slope protection for a new earth dam because the water from the river or stream is diverted in some manner around the construction site. Most soil-cement bank protection is constructed during dry seasons where there is no or hardly any flow in the waterway. However, unexpected rains or effluent from a sewage treatment plant can cause flowing water in the stream. In these cases, the contractor must either wait for the river to gradually recede or the water can be diverted using dikes placed adjacent to where the soil-cement will be placed. In some cases, water stored in a natural bank is released when the bank is excavated in preparation for placing the soil-cement protection. This water may be drained with a gravel drainage layer placed adjacent to the slope and below the first soil-cement layer. In order to properly construct the soil-cement traditional dewatering methods can also be employed. Section Dimensions

The most economical construction method of soil-cement bank protection is by using conventional highway construction equipment. This is because contractors already have the necessary equipment and do not have to make capital investments that can be employed only for a limited amount of work. This economic consideration manifests that the construction process determines the minimum thickness of protection. Compacted layer thicknesses have generally ranged from six to nine inches (150 to 230 mm) with eight to nine inches (200 to 230 mm) becoming more common on recent projects. Then, considering a minimum width of eight feet (2.4 m) needed for haul trucks with an eight-inch (200 mm) layer thickness, the minimum thickness measured perpendicular to the slope calculates to be about 5.2 feet for a 1.0H:1.0V slope and 3.9 feet for a 1.5H:1.0V slope of the bank protection. See Figure 2 for the typical section of a soil-cement bank protection constructed at Tucson, AZ.

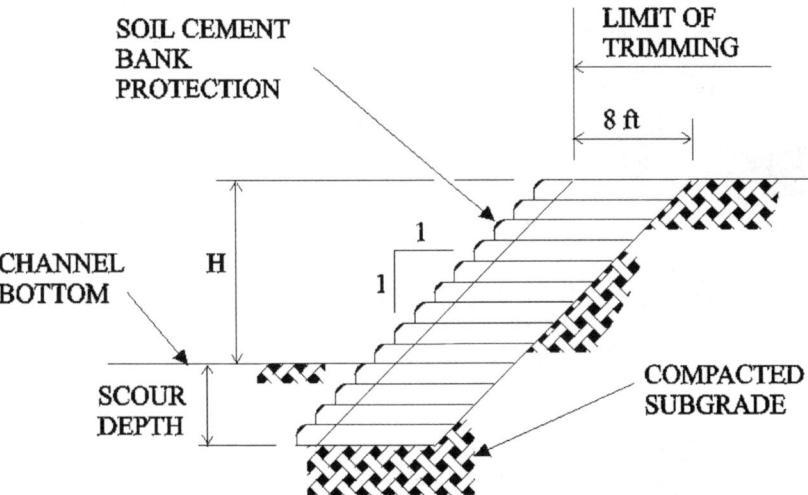

FIG. 2. Typical Bank Protection Section at Tucson, Arizona.

The elevation of the top of bank protection and the toe elevation can be computed from hydraulic and geomorphic considerations. The main factor in determining the crest elevation is the level of protection desired. This level is usually based on the 1 in 100 year flood, but some early designs in Tucson, AZ., were built to accommodate the 1 in 25 year flood. In the latter case when a flood occurred in excess of the 1 in 25 year event at Tucson in 1983, the soil-cement still performed well. When this higher flow occurred, the bank protection section was overtopped and there was erosion of soil from behind the soil-cement. Still, the erosion resistant soil-cement remained intact and in place. Following the flood, compacted fill was placed in the eroded areas and the bank protection was ready to withstand the next flood event.

Other factors to consider in establishing the elevation for the top of section are freeboard and the influence of debris, water and sand movement, and super elevation at bends.

The toe elevation for the soil-cement bank protection should be based on the many scour components computed for a 100-year event. While it is relatively easy to extend soil-cement bank protection upward, it is obviously more difficult and costly to place the toe deeper.

MATERIALS AND MIXTURE PROPORTIONS

Soil-cement is generally produced by using on-site soils that are mixed with portland cement (Type II), and water. Sand or sand-gravel for soil-cement usually comes from the banks or sandbars in the river to be protected.

Materials

Well-graded sands or sand gravel mixtures require the least amount of cement to provide adequate durability to resist deterioration due to weather or river flows. The gradation of these bank materials can vary depending upon location, but little processing of the river-run material is required. At most sites, the only processing required is to scalp off oversize clods, roots, or rocks on a 1½" (38mm) screen. However, material from the Salt River at Phoenix contains many large cobbles. In such a case, all material greater than three inches (75 mm) needs to be screened out. Additionally clay or calcareous cemented balls greater than one inch (25 mm) in size should be screened out and wasted.

To assure that the mixture can be properly compacted a minimum of 40% of the soil-aggregate should pass the #4 (5 mm) sieve. The amount of fires used in soil-cement bank protection projects has ranged from 3% to 20%. Clays with a plasticity index (PI) greater than six are generally not allowed, as they require a greater amount of cement for equivalent durability. Additionally it is more difficult to mix the clay and cement.

Mixture Proportions

With minimally processed on-site soil, standard laboratory tests on mixtures of soil plus cement are used to determine the three fundamental requirements for durable soil-cement (1) proper moisture content (2) adequate cement content and (3) adequate density

1. Moisture Content – the "standard" Proctor moisture density test (ASTM D558) is used in the laboratory to determine optimum moisture content and maximum density for the soil-cement mixture. The results of the test are used during construction to determine the amount of water to be added to the soil and the target density for the compacted soil-cement mixture.

2. Cement Content – the amount of cement specified is the minimum required to produce a material that will resist volume changes produced by external

variations in moisture (wetting and drying) and temperature (freezing and thawing). This amount of cement can be determined either by utilizing the results of laboratory durability tests (wet-dry and/or freeze-thaw) or by achieving a minimum seven-day compressive strength that correlates to a high level of confidence that the compacted mixture will be durable (commonly 600 psi (4.14 mpa)). For special conditions such as those encountered at Phoenix, AZ where coarser river sediments have greater erosion potential, a higher minimum seven-day compressive strength is specified (750 psi (5.17 Mpa)). Actual cement contents used for soil-cement bank protection have ranged from 7 to 12% by dry weight of soil.

3. Density – High density directly correlates to high compressive strength and therefore high erosion resistance. A target density can be determined either in the laboratory using standard Proctor compactive effort or by a test strip using actual materials and equipment planned for use on the project. Then, some slightly lesser density, such as 98% of the target density is specified for construction.

CONSTRUCTION OPERATIONS

The basic construction process, after control of water and excavation of the slope to the toe elevation, consists of proportioning and mixing, transporting, spreading, compaction, finishing, and curing the soil-cement. Special construction features that may be involved with a project include treatment of the outer edges, bonding successive layers, and construction joints. As with most construction projects, proper selection and utilization of equipment is the key to success.

Proportioning and Mixing

Central mixing plants of the twin-shaft pugmill variety are commonly used for volumetric proportioning (as opposed to weigh batching) and mixing the soil-cement. These continuous mixing plants with rated capacities between 250 and 500 tons/hr are able to produce between 170 and 340 cy of soil-cement per hour. The plants are usually located atop the bank in an area adjacent to the sand aggregate source or where there is sufficient space for processing and stockpiling the soil. Location atop the bank insures that, the mixing plant will not be affected by flood flows in the river. However, some contractors have accepted the risk and set their plants in wide riverbeds such as the Salt River provides at Phoenix. Such an arrangement offers easy access to the work and an abundant supply of stabilization material.

Transporting

Plant mixed soil-cement is transported from the mixing plant either by rear dump trucks or by bottom dump trucks. When the latter larger capacity truck/trailer piece of equipment is used, a width greater than eight feet (2.4 m) may be required to accommodate these larger and wider trucks. In some cases, the contractor, at his expense, may elect to build a wider soil-cement section in order to achieve a higher production rate. Access to the placement area from a plant atop the bank is usually accomplished by sloping earth ramps. Soil-cement ramps are sometimes part of the design to allow egress in and out of the river bottom – mainly during a flood event.

On some projects, where the design provided for a narrower placement width or where the access situation precluded the use of truck transport, conveyor belts have been used to transport the soil-cement, Figure 3.

FIG. 3. Conveyor belt feeds soil-cement to spreader box, Rio Puerco, Gallup, New Mexico.

Spreading

When dump trucks are used, spreading of the soil-cement to a uniform loose thickness for compaction may involve the use of a dozer pushed spreader box. However, dozers or motor graders (Figure 4) are also used to spread the soil-cement mixture when it is placed in a "ribbon" by a bottom dump truck. Some contractors have used an 8 ft. (2.4 m) wide spreader attached to the rear of a motor grader with good results, in this latter case.

Placement of soil-cement lifts should be limited to a total height of 4 ft (1.2 m) in a single 8 to 10 hour construction shift. This limit helps prevent possible stability problems. Instability of the layered soil-cement mass as evidenced by bulging of the outer face can occur if too much soil-cement is placed before the lower layers have attained sufficient stability to properly support the weight of the material and equipment above.

Compaction

To achieve density a variety of wheeled rollers have been used to compact soil-cement. In the early development of soil-cement bank protection, rubber-tired (pneumatic) rollers were used. In recent years, steel wheel (single or double drum), Figure 4, or pad foot vibratory rollers have been used to compact soil-cement mixes that used granular soils. At times the vibratory roller creates narrow, closely spaced cracks transverse to the direction of travel. A rubber-tired roller may be required for final compaction to close these cracks or striations. Any piece of equipment that consistently produces the specified level of density should be allowed by the specifications.

FIG. 4. Steel wheel vibratory rollers compact soil-cement along Salt River at Phoenix, Arizona.

Finishing and Curing

If the depth for any compacted lift of soil-cement exceeds specification, the use of a grader may be required to trim the soil-cement surface. While this problem is better controlled in the spreading operation, any trimming of the compacted soil-cement surface should be accomplished soon after compaction, as the soil-cement hardens quickly. Soil-cement surfaces need to be kept clean and continuously moist until the next layer is spread. However, the surface should not be puddled with water.

Curing of the top surface and the outer exposed face of the soil-cement should be for a minimum of seven days and preferably 14 days. Curing by a continuous spray of water is best, but a moist earth blanket at least 6 in. (150 mm) thick can be used for the top surface. Concrete curing and bituminous sealing membranes are seldom effective due to the low water content in the soil-cement mixture. Cost and aesthetics are other factors that may rule out membranes to cure soil-cement bank protection.

Weather Considerations

Soil-cement should not be placed in cold weather when freezing temperatures are expected at night. Most specifications in the southwest require the air temperature to be at least 45°F (7°C) or 40°F (4°C) and rising at the time of placement. Where freezing temperatures are anticipated at night, previously placed soil-cement needs to be protected against disruption to cement hydration caused by freezing. Adequate protection to maintain the temperature of the soil-cement above freezing can be achieved by covering the soil-cement with insulating blankets. Other protective methods, such as earth cover or straw, may be effective on the top surface, but cannot properly protect the sides of the steep soil-cement section.

SPECIAL CONSTRUCTION FEATURES

Construction joints

When soil-cement construction is interrupted, such as at the end of each day's placement or due to equipment breakdown, a full depth vertical construction-joint should be made. The vertical joint can be produced by cutting back into the fully compacted soil-cement lift to produce a vertical face perpendicular to the direction of layer placement. The loose material so cut should be wasted.

Bonding

Where there is a delay in placing the next layer within a specified time period or for some special design condition, bonding of successive layers may be required. The exposed layer should be kept clean and moist until grout or dry cement for bonding is applied. In either case, the bonding agent is applied just prior to placing the next layer of soil-cement and the preceding layer is not allowed to dry out. When dry cement is used for bonding, it should be applied at the rate of approximately 1 lb/sq.yd. (0.5 kg/m^2). The dry cement can be sprinkled with water prior to placing the next layer or activated by the water contained in the soil-cement mixture in the layer above.

One area where bonding is critical is between the uppermost two or three lifts for a soil-cement grade-control structure (drop structure). In this manner, improved shear values are obtained by achieving greater cohesion in an area where there is little weight above to mobilize shear-friction resistance.

Outer edges

The outer edges of soil-cement bank protection can either be:

1. Left ragged as is.

2. Trimmed to form a smooth slope, Figure 5.

3. Trimmed or compacted to produce definite stair-steps, Figure 6.

A smooth plane improves hydraulic properties during a flood. A wing blade attached to a dozer accomplishes the trimming. This should be done soon after final compaction. Waiting more than one day makes the trimming of hardened soil-cement more difficult.

FIG. 5. Smooth trimmed soil-cement bank protection, Rillito River at Tucson, Arizona.

FIG. 6.   Stair-stepped soil-cement bank protection, Sand Creek, Aurora, Colorado.

Definite stair-steps are desired in some areas to allow people and animals to climb out from the river bottom during flash floods. Stepped edges, that can be quite attractive, are invariably associated with slopes flatter than 1.0 H: 1.0.V. Contractors have devised many methods to cut, form or compact the outer edge of a soil-cement bank to produce a stair-stepped effect. It is desirable to have the exposed outer edge well compacted to a high density. This is an area where durability is most needed. Figure 7 shows a grade control structure on a small waterway near Albuquerque, NM. Definite steps were produced by overbuilding and cutting back to form an aesthetically pleasing vertical surface.

FIG. 7. Cut steps on soil-cement drop structure on La Barranca Arroyo near Albuquerque, New Mexico.

COST

The in-place cost of soil-cement bank protection can be divided into two components: 1) the cost of cement and 2) the cost of producing the soil-cement. The cost of cement includes the delivered cost of the portland cement, plus the cost of on-site storage. The costs associated with producing the soil-cement basically consist of all construction related items except the cement. These include the sand aggregate as well as proportioning, mixing, transporting, spreading, compacting, trimming and curing the soil-cement.

The cost of cement is hardly affected by the size of the project (i.e. volume of soil-cement). However, the cost of producing soil-cement is quite volume dependent and involves many factors. Invariably, the cost of producing soil-cement decreases as the volume to be placed increases. Factors affecting the production cost include:

1. Availability and haul of aggregate,

2. Ease of construction,

3. Specifications, and

4. Competition.

Obviously, low bid prices for soil-cement are obtained when aggregate is located on-site and requires little haul or processing. Other factors tending to reduce cost are: good access between the mixing plant and placement area, little need to control water at the site, projects having large volumes of soil-cement and a reasonable number of experienced bidders.

A study of the low bids received on a number of soil-cement bank protection projects received between 1994 and 1997 indicates the bid price for cement was generally between 40-60% of the total in-place soil-cement cost.

The cost of producing soil-cement varies inversely with the volume of material to be placed. As expected, larger volumes are bid at lower unit costs. Therefore, for larger volumes of soil-cement, the cost of cement tends to be a greater percentage of the total in-place cost and viceversa.

Table 1 shows the average of low bids received for construction of soil-cement bank protection between 1994 and 1997. Most of these bids were projects located in Arizona. It is not known if any of the projects considered were unbalanced or some of the soil-cement cost was placed in a pay item earlier in the construction sequence.

Included in this cost information supplied by the Portland Cement Association, the cost of cement for some large volume projects was obviously unbalanced as cement was bid considerably below cost. Also, some small volume projects were bid as a single unit price for soil-cement rather than two bid items (cy of soil-cement and cwt of portland cement.) Both of these situations were not considered in Table 1. Bidding soil-cement as cement plus production cost is recommended. In this way, the cement content can be varied if the pit-run soil aggregate conditions change. Also, the contractor will get paid for all cement that he actually uses within the pay lines of the section.

Table 1. Bid Prices for Soil-Cement Bank Protection 1994-1997

Year Bid (1)	No. of Projects (2)	Ave. vol. cy (3)	Ave. lb cement/cy (4)	Cost cement/cwt (5)	Ave. cost of cement/cy (6)	Total cost/cy (7)	% cost Cement/ Total (8)
1994	8	27,300	337	$3.94	$13.26	$23.97	55.3
1995	10	17,100	331	$3.88	$12.85	$25.80	49.8
1996	4	76,300	299	$4.37	$13.06	$26.29	49.7
1997	2	16,300	299	$4.13	$12.31	$26.32	46.8
4 yr. Weighted Ave.		30,300	325	$4.00	$13.00	$25.32	51.3

From the table, it shows that the weighted average pounds of cement specified for soil-cement bank protection was 325 lb/cy. This table can then be used for estimating future projects. Because the cost of cement changes with demand, the next step is procuring a delivered price for cement at the site from a cement producer. A certain percent for handling cement (usually less than 10%) can be added to the supplier's quote. Then, the resulting cement price/cwt (hundred weight) can be multiplied by 325 lb/cy. (or the project specific cement content) to determine the cement cost per cy of soil-cement. With cement being between 40-60% of the total cost, the total cost can be estimated. If 50% is used, this means the total cost of the soil-cement should be double the cement cost previously determined. For initial estimating purposes, the amount probably should be increased by 15% as all the data was based on the price submitted by the low bidder.

This estimate should then be tempered by any conditions that may tend to increase or decrease construction cost. If there are no obstacles to soil-cement placement an experienced contractor should be able to place about 200 cy per hour on a large volume project.

CONCLUSIONS

Stream bank erosion is a severe problem when flash floods occur in the arid southwestern United States. Cement-stabilized on-site sand is an economical solution for addressing this problem. The key to an economical design is careful consideration of the factors that affect the soil-cement placement rate; and it should be remembered that the plant rarely controls the production rate. Typically the most economical design is one that permits the use of conventional highway construction equipment to construct the soil-cement bank protection. Therefore, it is best to let the construction process determine the minimum thickness of protection.

The acceptance of soil-cement as a bank protection method is attributable to its low cost, ease and speed of construction, and its proven performance through many flood flows. Projects have even been subjected to overtopping with no appreciable loss in structural performance.

APPENDIX. REFERENCES

1. Hansen, Kenneth D. and Lynch, John B., (1995) "Controlling Floods in the Desert with Soil-Cement." Second CANMET/ACI International Symposium on Advances in Concrete Technology, Las Vegas, NV, June.

1. Portland Cement Association, (1997) "Guide for Estimating Cost of Soil-Cement Slope Protection for Construction in 1997." Denver, CO, March.

QUALITY CONTROL OF SOIL-CEMENT CONSTRUCTION FOR WATER RESOURCES

Randall P. Bass, P.E. Member[1]

ABSTRACT

The uses of soil-cement in the water resources applications have broadened over the past years. Still one construction related task has remained essentially the same: quality control. Soil-cement is a highly compacted mixture of soil; usually sand, portland cement, and water. It has the properties of a low-strength concrete. Without adequate controls on the quality of the materials and construction methods, the performance of the project may not meet expectations.

A successful soil-cement project begins with the development of a quality control program with the responsibilities of each party clearly defined. The quality control plan should address submittals which must receive approval from the project engineer; mixture proportioning considerations; minimum requirements for equipment such as the mixing plant, compactors, and curing equipment; types and frequencies of material testing; and acceptance criteria for construction procedures and individual material tests.

Quality control practices during construction do not have to negatively impact the contractor's production. Each member of the project team, from the laborer to the project manager, can be aware and trained in their responsibilities to ensure a quality product. The benefits of having everyone on the project buy into and execute their responsibilities will minimize any rework and ensure that the performance criteria will be met or exceeded.

[1] Program Manager Water Resources, Portland Cement Association, 517 Haralson Drive, Lilburn, GA 30047

This paper will focus on quality control issues where soil-cement is used for bank stabilization grade control structures and slope protection. Issues involving mixing of soil-cement will be directed at central plant pugmills.

INTRODUCTION

The early work on soil-cement performed by the Bureau of Reclamation started with the 10 year durability study in the early 50's at Bonny Dam in Colorado. Soil-cement was used as slope protection and subjected to numerous freeze/thaw and wet/dry cycles. The success of this study led to the exponential growth of soil-cement for slope protection and spillways in the 60's and 70's.

Engineers soon realized that this technology could be transferred to stabilizing stream banks to protect against erosion. The use of soil-cement for bank stabilization applications has increased tremendously over the past 20 years due in large part to its proven performance during significant flood events. The stream bank stabilization projects in Phoenix and Tucson, Arizona and in Orange County, California have experienced numerous flood events with little if any damage requiring anything other than minor maintenance work. Projects are now being designed and constructed where channel gradients produce flow velocities over 6m/sec (20 ft/sec). The success of using soil-cement for bank stabilization can be credited to well developed designs and to a detailed quality control program in place during construction to ensure that the materials and the results of the methods of construction met the design requirements.

The first step in developing a successful quality control program is to tailor the program to the needs of the specific project and develop a clear understanding of the responsibilities of each member of the project team. The type of tests, testing procedures, frequency, and reporting protocol must be covered in detail in the construction documents and discussed during pre-construction meetings. In addition, a schedule of submittals must be developed which identifies sources and types of materials, proposed equipment, and work schedule so that the appropriate resources can be allocated for quality control activities.

Typical pre-construction submittals should include as a minimum the following:

- The proposed source(s) of soil aggregate and type and source of cement to be used in the soil-cement.
- The proposed size and number of soil aggregate stockpiles.
- The mixture proportions to be used to conform with specifications.
- Type of mixing plant proposed with information on capacity, mixture proportioning accuracy, and the controls for monitoring cement, water and aggregate feed.

- The approximate length of soil-cement to be placed prior to starting compaction operations.
- The type of transporting, spreading, curing, and compaction equipment to be used.
- The methods to be used to keep surfaces continually moist until subsequent layers of soil-cement are placed.
- The method to be used to cure permanently exposed soil-cement surfaces.

The execution of a quality control program during construction needs to be consistent but flexible enough to adjust to changes in the contractor's schedule and /or means and methods of construction. Many of the tasks that are a part of a quality control program are discussed below.

DEVELOPMENT OF MIX DESIGN

The economical savings of designing with soil-cement is the result of two factors. The first factor is the use of on-site soils as the aggregate for the soil-cement. If the on-site soils can be used with little processing, then the only material brought on site will be the cement. Secondly, because time is money to a project, it is advantageous for the contractor to develop a means and methods and construction sequence which allows the soil-cement to be placed in the shortest time.

The basic mix design principle with soil-cement is to utilize in-situ soils with minimal processing and optimize the cementitious content which will produce the specified performance measures. Once the desired proportions of cement to soil aggregate are set then achieving the maximum density and optimum moisture content will determine the ultimate strength and durability properties.

The most commonly used test to determine optimum moisture content and maximum density is the "standard" Proctor density test (ASTM D-558) for both molding laboratory test specimens and for field control during construction. The "modified" Proctor (ASTM 1557) has been used and is the more appropriate test when a significant amount of material is retained on the 4.75 mm sieve (No. 4) (greater then 30%). During construction, density tests should be performed periodically because the gradation and moisture content of soil-cement stockpiles can vary and adjustments have to be made to the target density the contractor must achieve.

After the optimum proportions of the soil-cement ingredients are determined in the laboratory, it is not uncommon for changes to be made to the amount of cementitious material and /or water content based on the results of the early strength tests of the initial field samples. Changes made to either cementitious content or moisture should be done in small incremental steps and held at one level until enough strength tests can be completed to determine the effects of the adjustments to the mix design. As part of the on-site quality control program, a detailed data base of all tests results and adjustments to the mix design should be maintained and analyzed so trends can be identified quickly and further adjustments made if needed.

STOCKPILE DEVELOPMENT

The development of the soil-cement mix design in the laboratory is based on a small sample of soil typically taken from a test pit in the area of the anticipated borrow source. This sample will generally not be representative of all the materials that will be encountered in the borrow area. River deposited soils can be vertically and horizontally stratified with lenses of course gravels, clean sands, silts, and clays.

The best soils for soil-cement are well graded with approximately 25 to 50% retained on the 4.75 mm sieve (No. 4), thus 50-75% passing the 4.75 mm sieve (No. 4) and 5-20 passing the 0.075 mm sieve (No. 200). Any material greater than 7.6 cm (3 inches) in size should be scalped from the stockpile.

The best method of constructing a stockpile when the borrow source is stratified with an assortment of different soils is to develop it in horizontal layers with varying types of the in-situ soils. During soil-cement production the loader which feeds the plant takes a full vertical scoop from the stockpile to obtain portions of the various types of soils. It is good practice to run just the soil aggregate through the mixing plant without adding any cement and perform a gradation test of the blended soil before soil-cement production begins. This gradation should be compared to the gradation of the samples used to develop the mix design in the laboratory. If significant differences are observed, new soil-cement test specimens should be made with the soil taken from the plant to see if the cementitious content needs to be adjusted before production begins.

A project in Santa Clarita, California demonstrated the problems that can occur with developing mixture proportions in the laboratory when the in-situ borrow soils vary significantly. Two independent geotechnical firms reported that the borrow source materials had significantly different gradations from one another. One reported an average of 46% of the material would pass the 0.075 mm sieve (No. 200) while the other reported less than 10% passing the 0.075 mm sieve (No. 200). Depending on the location of each of the samples, both were correct. Soil-cement laboratory tests showed that the cement demand to achieve the design strength ranged from 8–12% by weight of dry aggregate because of the differences in the reported gradations. After construction of the stockpile and running the soils through the plant, the range of materials passing the 0.075 mm sieve (No. 200) was between 8 and 18 percent. The 7-day strengths were achieved by using a cement content of 9%.

Continuous monitoring of the gradation and moisture content of the aggregate used in the soil-cement is necessary so timely adjustments can be made to the mix proportions. The quality control program should require several gradation tests a day until it can be established that the material is fairly uniform. Then the frequency can be reduced and a moisture content determined prior to start of production each day and again at midday.

PRODUCTION CHECKLIST

At the beginning of each day several tasks should be completed before any soil-cement is produced. As with conventional concrete soil-cement can be harmed by freezing temperatures. If construction occurs in a cold climate the air temperature should be measured to see that it is above 7 C degrees (45 F) or 4.5 C degrees (40 F) and rising. Next the condition of the previous day's lift surface must be inspected and approved. The lift must be clean, scarified if required, and moist. If a bonding mortar is required at cold joints the materials and labor must be on-site. The above controls must be approved before production for the day can begin.

Mixing

Plant mixed soil-cement can provide a uniform product at high production rates if proper controls are in place. Prior to production the plant must be calibrated to ensure that the various components of the mix are properly proportioned. Separate plant calibrations should be performed at different production rates that bracket the anticipated range of production rates that may be used during construction. During construction it is advisable to reconfirm the plant's calibration periodically to insure its reliability. The experience of the plant's operator can be an important asset in maintaining a proper proportion mix. By observing the color of the mix any abrupt changes in cement content may be detected and by feeling the material the moisture content can be monitored for changes.

After the plant is calibrated the next task is to check whether the plant can produce a product that is uniformly mixed on a consistent basis. This is referred to as a mixer uniformity test. The plant is operated at its anticipated production rate and samples of the soil-cement are taken at different intervals. A series of tests are performed such as gradation, compressive tests, water content, and unit weight. The results of these tests are compared to determine if the mixing plant is producing a uniform product. If the results of these tests are outside acceptable limits, modifications to the plant are required. Modifications may include increasing the mixing time, changing worn paddles, or calibrating weigh scales to name a few.

If the pugmill proposed for the project has successfully produced soil-cement on a recent project and the operator has significant experience with the plant the uniformity tests may be unnecessary. If results of strength tests show a wide range of results then the uniformity tests should be considered. If drum mixing is proposed, the uniformity tests should be performed before production commences. A typical continuous pugmill is shown in Figure 1.

Cement Content Determination

As a gross check of the cement content of the soil-cement placed during a particular shift, a review should be made of the cement delivered to the job site versus the amount of cement in storage onsite to determine what was theoretically used in the soil-cement produced during that shift. This quick rough check can identify any significant cement metering problems with the plant.

If concerns over cement content becomes an issue then the ASTM D-5982 Heat of Neutralization test should be performed. This test can be performed quickly without a lot of specialized equipment. The test can produce accurate results in soil-cement samples.

Conveyance and Placement

The most common method of transporting the soil-cement from the mixing plant to the placement area is by end or bottom dump trucks. Following the deposit of the soil-cement onto the placement surface a motor grader, small dozer, or spreader box will spread the soil-cement to the proper width and thickness. However, before any soil-cement is placed attention must first be focused on the subgrade. The subgrade should be compacted to at least 95% of standard maximum density and proof rolled to identify any soft areas. Just prior to placing soil-cement the subgrade should be moistened. If soil-cement is to be placed on previously placed and compacted soil-cement, no further density tests or proof rolling is required.

The quality control practices that need to occur during this operation are simple but important. Because trucks are operating on and off the lift, surface contamination of the lift surface may occur. Preplanning and then monitoring of the traffic pattern and haul routes should take place on a continuous basic. After rain haul routes should be checked prior to commencing soil-cement operations. Some projects have deep cuts for scour protection in a stream bed in which backfill material is placed concurrently with the soil-cement. This operation is conducive to contaminating the soil-cement lifts if special care is not taken.

During the spreading of the soil-cement the loose thickness of the material must be checked at regular intervals. If the soil-cement is placed too thick, adequate compaction near the bottom of the lift may not be achieved. Layers placed too thin have a tendency to spall and unravel. A probe rod with an indicator strip marking the appropriate depth can be used to monitor lift thickness. Any adjustments to the thickness of a lift should be made prior to any compaction to the specified compacted thickness.

Another control element that must be monitored during placement operations is the width of each lift. Adequate survey controls must be available so offsets can be easily measured to set the outer edge of the bank protection or drop structures. It is difficult to regain the desired outer slope edge when the preceding lift is placed too narrow.

Compaction

The compressive strength and durability of soil-cement has a direct relationship with the density achieved in the field. Typically, in the field the criteria for density acceptance will be a running average of a certain number of consecutive in-place density tests not less than 98% of the maximum density obtained by ASTM D-558 with no individual test less than 95%. Field density tests are performed in accordance with AASHTO T 238 "Nuclear Method" (Figure 2) or with ASTM D-1556 "Sand Cone Method". The nuclear gauge method is most commonly used with the sand cone method used as a calibration check occasionally.

Specifications will place a time limitation when water is added at the mix plant to when final compaction is achieved. Typically one hour is allowed. This time to compaction is important for the ultimate strength gain of the soil-cement. If too much time passes the cement paste starts its initial set and if the soil-cement is then compacted some of the cement will not reset. This time restriction may be difficult to track during construction if different loads of soil-cement are going to different placement locations. Knowing when each truck load was mixed is almost impossible to track. The quality control personnel must develop a general feel for time and every so often actually time a complete sequence.

The method of compacting soil-cement can be by vibratory smooth drum (Figure 3) or pad foot rollers or by pneumatic rubber tire compactors. At the beginning of each day the equipment should be checked for fluid leaks and periodically have the frequency of vibratory roller spot checked. Before the start of most projects a test strip of soil-cement is placed to determined the number of roller passes necessary to achieved the target density. During compaction operations, the procedures should be monitored for the correct number of passes, overlapping lanes, and whether general good rolling practices are followed. This becomes more important when a new operator is placed on the roller.

During production the number of density/moisture tests performed is based on time and/or number of cubic yards placed since the last test. A typical specification will require a test at least every two hours but not more than once every 500 cubic meters (650 cu yards). If the results of the tests are below the minimum allowed the area can be re-rolled if the time to compaction has not been exceeded. A retest of the density should follow the re-rolling. On hot windy days enough moisture can be lost from the soil-cement surface prior to compaction so that achieving the density target may be extremely difficult. Under these types of climatic conditions additional water should be added at the plant to compensate for the moisture lost to evaporation. Periodically, after determining the moisture content of the compacted soil-cement by the nuclear gauge in the direct transmission mode the same soil-cement should be taken to the lab and oven dried to check the accuracy of the nuclear gauge.

Curing

Temporarily exposed compacted soil-cement surfaces must be kept moist until the subsequent layer is placed on top. Too much or too little water can negatively affect the ultimate strength of the soil-cement. Water should be applied at a rate which will moisten the surface of the compacted surface without leaving puddles. Water trucks are typically used but before they are allowed to operate on the soil-cement surface they must demonstrate they can apply a light spray the full width of the lift surface. Heavy applications of water can create a thin layer of soil-cement at the surface that will either have a high water to cement ratio or be predominantly sand because the cement was washed out. No soil-cement surface should be allowed to dry out completely. Improper curing will result in low strength soil-cement. If curing of the soil-cement is inadequate, removal of the soil-cement may become necessary.

Permanently exposed soil-cement surfaces must be kept moist for a minimum of seven (7) days. This curing period may be longer if the project is constructed in a hot and dry environment. Records must be kept of the completion dates of each section of the project so an appropriate curing schedule can be created and monitored. If atmospheric conditions may produce temperatures below 0 C degrees (32 F), the soil-cement must be protected from freezing until it has obtained its seven day age.

Joints and Finishing

Joints are quite common in soil-cement construction. Transverse construction joints are formed at the end of each day's work or when operations are interrupted for more than a few hours. The joint is created by cutting back into the completed lift a short distance from where that lift stopped for the day to form a full-depth near vertical face. The station and lift elevation should be recorded of all joints. Transverse lift joints should be staggered so that a series of joints do not line up over one another.

Horizontal cold joints are created when there is an interruption in soil-cement placement from one lift to another that exceeds the time specified in the specifications for the development of cold joint. The location and time of day must be recorded of each cold joint so the appropriate method of cold joint treatment can be applied. Usually the type of cold joint treatments will vary if the soil-cement surface is exposed from two to eight hours. If a cold joint develops prior to restarting soil-cement placement in that area the type of joint treatment must be decided and the lift surface approved.

Typical lift treatments include sweeping the surface with a power broom and scoring the surface to create a rough surface to increase shear resistance with the next lift. If cohesion is important between cold joints a bonding grout or mortar will be specified to "glue" the lifts together. This bonding material can either be a premixed high slump cement grout or mortar or possibly dry cement placed over the lift and watered in-place. Monitoring of lift joint treatments is imperative to ensure each soil-cement mass acts as a single monolift and not individual layers. Field records should be kept of the location and type of each joint treatment.

Soil-Cement Strength Testing

During soil-cement placement, samples should be made for strength testing in accordance with the specifying agency prescribed methods. Typical procedures consist of taking a sample of soil-cement from the placement area (Figure 4) and compacting the material into Proctor molds in accordance with ASTM D-558 (Figure 5). The samples are then extracted from the molds and cured until the specified maturity. The frequency of sample preparation is usually based on the number of cubic yards placed. The age at compressive strength testing can vary from 1 to 28 days. The strength criteria is based on the strength at seven days but some agencies will also test at earlier ages to help detect mix design problems before a lot of soil-cement has been placed. It is also a good idea to keep extra samples for 28 day breaks in case the seven day results do not meet the minimum strength requirements. The reporting and analysis of the test results should be prompt so any changes needed to the mix proportions can be incorporated in a timely manner.

CONCLUSION

During the initial days of soil-cement placement more frequent testing and close monitoring of construction practices is required until there is little variance in the test results and the construction personnel is well versed in the proper methods of working with soil-cement. Prompt reporting of all tests results to the responsible parties will expedite any adjustments to the mix proportions and/or construction practices.

Quality control is everyone's business during construction. If at the beginning of the project, everyone is properly educated in all the quality control requirements and the significance of those requirements, the contractor should be able to achieve high production rates without negatively impacting the performance of the project.

REFERENCES

Hansen, K.D. and Lynch, J. B., Controlling Floods in the Desert with Soil-Cement, Second CANMET/ACI International Symposium on Advances in Concrete Technology, Las Vegas, NV. June 11-14, 1995.

Soil-Cement for Water Control, Portland Cement Association, 1996, IS 166.02W.

Soil-Cement Laboratory Handbook, 1992, Portland Cement Association, EB 052.07S.

Figure 1. Typical Continuous Pugmill in Operation

Figure 2. Density Control with Nuclear Gauge

Figure 3. Compaction with a Smooth Drum Vibratory Roller

Figure 4. Sampling Soil-Cement from Placement Area

Figure 5. Preparing Soil-Cement Samples in
Accordance with ASTM D-558.

Soil Stabilization/Soil Cement Mark-Lang, Inc.'s Approach

By William F. Boswell[1]

Abstract: Bidding soil cement is somewhat unique. The nature of mix in-place soil cement does not allow a 7 a.m. to 3 p.m. work schedule. When the cement is applied the material cannot be left overnight, you must complete the operation even if overtime work is necessary. Working around utilities can be a very challenging matter. Of major concern during a soil cement operation is the amount of rock in excess of softball size. Successful completion of a project requires the use of proper equipment. It is important to determine and use the appropriate pulvermixer for the soil type.

INTRODUCTION

Bidding soil cement is somewhat unique. There are many things that need to be assessed. Timing is very important both on a daily and on a seasonal basis as temperature plays a critical part in the performance of soil cement. The nature of mix in-place soil cement does not allow a 7 a.m. to 3 p.m. work schedule. When the cement is applied the material cannot be left overnight, you must complete the operation even if overtime work is necessary.

The Project Site

When bidding soil cement the configuration of the areas to be treated must be considered. There are many aspects of the process that can be affected by configuration. If a delivery truck has to wait the delay can effect cost. The proposed areas to be treated should be checked for tight cul-de-sacs, and availability of a water source. It maybe necessary to travel over treated areas to get water to treat other areas of the project. In some instances sites are small and it is very difficult to load and unload equipment in a safe manner. In the case of a small development with only one access road it can be difficult to deliver equipment and cement to the project.

---

[1] Secretary/Treasurer Mark-Lang, Inc., P. O. Box 322, Millersville, Maryland 21108, E-mail: marklang@starpower.net

Other trades working in a development can cause unforeseen problems. Parking for workers can be a problem. Deliveries of building materials during operations can cause stress, tension, and reduced productivity. This is particularly true when an operation has to stop for delivery of building materials to the tradesman who just had to move his truck from the street that is about to be treated.

Dust

Cement and/or lime dust is another issue that should be taken into consideration. Whether treating a commercial lot, or working in a new or expanding subdivision cement and/or lime dust can cause damage to nearby vehicles. Advise clients of this potential, have them advise their employees and other tradesmen who may be in the area of the possible damage. Always work with favorable wind conditions.

Utilities

Working around utilities can be a very challenging matter. A lot of handwork can go into working around a manhole that is especially close to a section of curb. Removing the surrounding material with a Bobcat, treating it in adjacent areas and replacing the treated material again with the Bobcat is an effective technique for handling tight situations. Hand working an area is very labor intensive and costly. If this situation cannot be avoided dollars must be added to the bid.

Protection of utilities is a very important way to avoid unnecessary back charges. When possible the water value box should be removed prior to mixing and replaced after finish grading. If the General Contractor/owner will arrange for the removal and replacement this can be very helpful and time saving if properly coordinated. Special attention should be given to the treatment of utility trench lines. Many times proper density in not achieved when material is removed and replaced in a utility trench. Discuss with the General Contractor/owner plans to treat these areas if they do not pass a proof-rolling test. Additional depth in such areas is a good option. If the General Contractor and/or owner does not agree to additional treatment be prepared for back charges for failed soil cement in trench line areas. A camera can be a very valuable tool in those situations.

Any soft areas on the site should be identified and a plan of resolution discussed with the client. Inlets can be another back charge nightmare. Take precautions to protect the structures and eliminate material going into the inlet. If treated material falls into the inlet it should be removed immediately, before it hardens.

Be sure to call the Utility locator service to insure that all hidden lines will be identified. The gas or power line was probably 36 inches deep when it was installed, but now the site contractor has moved material and the line could be just below the surface. New construction that is tying into accessing streets can cause a similar problem. Utilities are in, but may not be active this is something that can be misleading. Damage to these utilities can be costly.

Rock

Of major concern in a soil cement operation is the amount of rock in excess of softball size. Rock larger that that can damage the pulverizing equipment and make grading difficult. Removal of oversize rock is labor intensive, making it a cost consideration. A site that is on grade but which still has a large amount of rock to be removed may end up low, due to the removal of so much volume. Discussing with clients the potential of swell from soil cement process allows the pre-grade operation to make the necessary adjustments. Making the client aware of the impact of unacceptable rock in the grade results in better efforts to remove it during site grading.

Mix Design

The mix design prepared by the Engineer is of the utmost importance. When bidding try to be very clear on how the application rate is expressed, by volume or weight. Always double check the depth of mix and address any other concerns. Checking with the Engineer allows all parties to be clear on the anticipated outcome of the treatment.

CONSTRUCTING THE PROJECT

All of the above are reviewed when preparing to move on to the job site. A site check is done just prior to arrival. The grade is spot-checked. Visually assess the amount of rock in the grade. Any obvious soft or yielding areas are noted. If conditions warrant proof roll the entire site. Make notes of any existing curb and gutter damage. Some developers are quick to find anyone grading at fault for curb and gutter damage. It is very difficult to grade along the edge of curb and gutter when the curb and gutter is not properly installed or not fully cured. This can result in costly misdirected back charges. Again a picture is worth a thousand words. The final pre-arrival step is to determine the yardage prepared for treatment. This is required to order the necessary amount of product for the first day. Watch yield calculations on a day to day basis. To avoid a short load at the end of the job it maybe necessary to adjust an order. Short loads can generate minimum freight charges.

Scarifying

Arrival on-site begins with scarifying the area to be treated that day. Scarifying, loosening the soil, can help identify problem areas. If rocks are an unforeseen problem scarifying can bring the unacceptable rocks to the surface. Removal of rock is no ones favorite job and being a "Rock Star" is not the most desired position on the crew. However, a good effort to remove rocks will save equipment from damage. If there are excessive amounts of rock to be removed, check the grade again, document, and discuss concerns with the client. Scarifying can also reveal soft, yielding, or wet areas and hidden organic matter.

Spreading

During the spreading operation observe closely the areas at utility structures and trench lines. The bulk application (spread rate) is checked and any adjustments are made. The calibration of equipment is a crucial point in federal, state, and airport work. These checks may need to be confirmed with the inspector or site manager.

The spread pattern of the product is determined by site configuration. Our bulker types vary somewhat and the size of the site will determine the bulk spreader to be used. The bulker should move at a continuous rate without stopping while the product is being discharged, Fig. 1.

FIG. 1. Bulker/spreading.

Mixing

Immediately after application of the product the first dry mixing is accomplished, Fig. 2. This mix should be checked to assure the uniformity of blend. It is important that the product be blended thoroughly with the soil in order to achieve the desired finished product.

The dry mix is followed by the application of water to bring the mix to optimum moisture content. In the case of very dry or windy conditions pre-wetting the grade can be helpful for getting enough moisture into the material and controlling dust. In extreme conditions pre-wetting prior to scarifying the material may be a necessity. During the application of water the water truck drivers should take care not to allow the discharge of water to puddle or pool. Such situations can result in soft or yielding grade conditions. These soft spots may in turn become areas where compaction cannot be achieved.

The second mix follows the application of water. Strict moisture control should be adhered to at all times. The control of moisture will help reduce reflective cracking. The normal mix procedure at Mark-Lang, Inc. is in a down cut motion using tine type mixer teeth, Fig. 3. In cohesive soils an up cut machine with conical type bits is better.

FIG. 2. Mixer.

In some areas it is required that the application of water be performed with the water being distributed through the mixer. This requires the water truck in front of or alongside the mixer. This is not the most practical method of operation.

During the mixing it is important that the operator watch for utilities and structures. The mixing operation should also be a continuous process with only the application of water between mixes. When working at inlets and manholes, after the wet mix the material is dug out deeper around the structure and additional product is added. Then the material is replaced and plate tamped. This creates a thickened edge at the structure. Again, caution is advised to avoid allowing treated material to enter the inlet or manhole.

Compaction

Initial compaction with a vibratory pad foot roller immediately follows the wet mix operation, Fig. 4. A roller pattern should be established at the beginning of the initial compaction effort. Normally two to three complete roller passes will be required. The depth of mix, type of soil, and required density are factors that control the number of roller passes required. The compactor should keep pace with the mixing operation. Special attention should be given to working around utility structures, and curb and gutter. The roller can cause costly damage to these structures. During compaction operations extra effort may also be required at trench lines.

Following the initial compaction the treated material should be shaped to the approximate line and grade. Special attention should always be given to grading for proper drainage. It is hard to make grade corrections to cured material. The compaction effort should continue until required density is achieved.

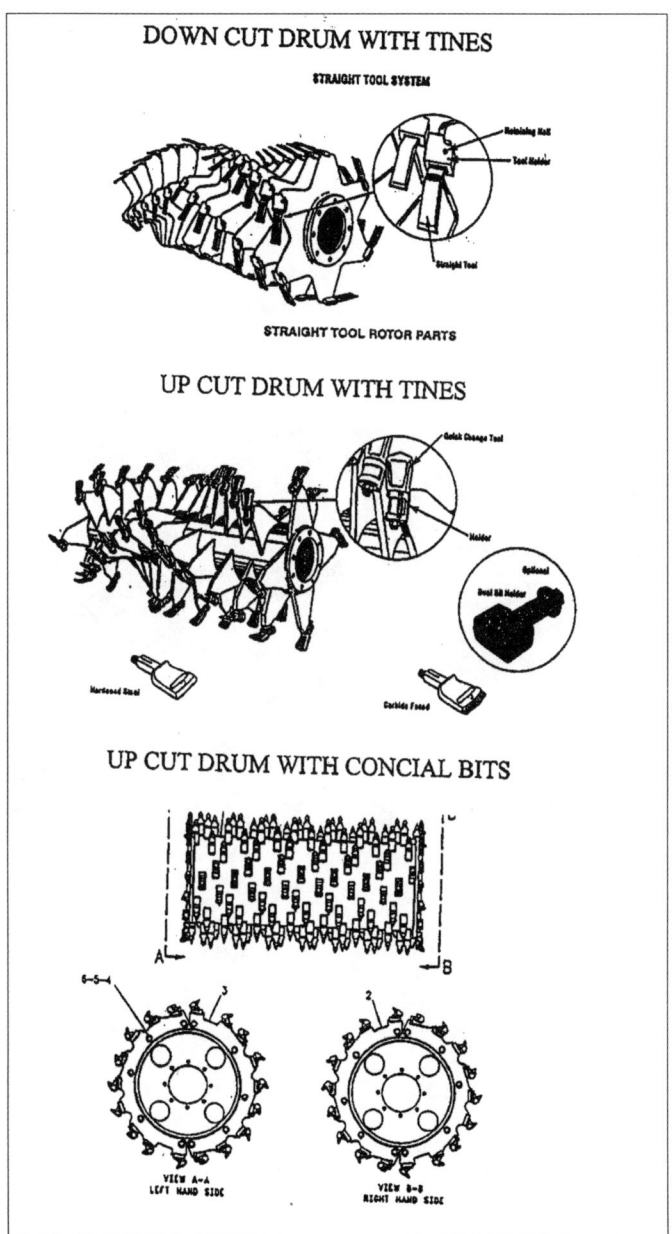

FIG. 3. Mixer Teeth, Tines and Bits.

The fine grade operation should follow acceptable compaction and rolling with a smooth drum vibratory or pneumatic roller, Fig. 5. The treated material should be kept from drying at all times during shaping, rolling and fine grading, Fig. 6. Again, the water truck drivers should exercise care not to allow puddling or pooling of water to occur.

FIG. 4. Initial compaction with a pad foot roller.

Curing

There are several acceptable methods of curing soil cement. Some projects will specify the use of liquid asphalt curing. That method has possible environmental impacts. The material can be tracked onto adjacent roadways requiring a clean up operation. In many areas the effect of run off is of great concern particularly for projects located near major waterways. The use of white curing compound has proven effective. It has to be applied at a heavy rate (0.25 to 0.30 gallons per square yard) in order to achieve complete coverage. This can be a time consuming operation and requires special equipment. Therefore there is a preference for the wet cure method, Fig. 7.

FIG. 5. Final compaction with smooth drum.

FIG. 6. Fine grading.

FIG. 7. Wet Curing.

Keep the finish grade wet/damp for a period of seven days, with as little traffic as possible being allowed to travel on the grade. The water truck should be careful not to make sharp turns during the initial curing. It is very important that the client understand the importance not allowing travel upon the treated area. Construction equipment particularly, track machines, should be strictly prohibited from entering the area.

During the curing operation the treated area must be kept from freezing for seven days. Frost should not be allowed to from on the grade during the first 48 hours of the curing process. The use of straw (4 inches minimum) and/or poly can help reduce possible freezing, depending on the projected low temperature. Temperature should always be taken into consideration at the beginning of a project. Do not begin a project without a weather projection that will allow for seven continuous days of temperatures above freezing after completion of mixing and grading. A thin layer of straw will help reduce the possibility of frosting.

CONCLUSIONS

Although all of the above points are very important to successful completion of a project, from our company's experience, the following are of the utmost importance.

- Determine and use the appropriate pulvermixer for the soil type. In clay soils an up cut machine with conical bits will result in a better initial mix.

- Identify soft yielding areas and trench lines. Repairs to these areas should be made well in advance of the application of product. Recently repaired areas will fail or breakdown again. Do not depend on a bridging action so solve trench line problems.

- It is very important to reach optimum moisture prior to spreading and mixing this is particularly so when treating sand, fine grained, or silty soils.

Mark-Lang strives to provide clients with the stabilized results they anticipated. This is accomplished by working with the Engineer, General Contractor, and the Owner to coordinate the work and provide a safe and productive work site. It is important to have a safe and productive work site both for Mark-Lang employees, and for other site personnel and contractors. We are very proud of several noteworthy and high productivity projects Mark-Lang, Inc. has completed.

Recycling Flexible Pavements with Cement:
Diverse Methods Produce Durable Pavements

Jan R. Prusinski, M.ASCE [1]

Abstract

Thousands of miles of existing flexible pavements throughout the U.S. are deteriorating. Street and highway departments continually seek economical, yet durable ways to rehabilitate their rutted and pot-holed roadways. Many have chosen to recycle their worn flexible pavements with cement. Recycled flexible pavements save money by utilizing in-situ materials, thus eliminating haul/disposal costs, and avoiding the expense of producing and importing new aggregate. Cement-treated flexible bases provide a moisture-resistant pavement base capable of supporting light to heavy traffic for many years. This paper discusses mixture and thickness design procedures, presents the basic construction sequence, details recycling costs and documents quality control/assurance issues. Projects in Texas, Washington, California and Florida are presented as examples which demonstrate the diverse methods used to produce durable pavements at minimum cost through recycling.

Introduction

As the nation's road system ages, the emphasis on road construction shifts from building new highways, roads and streets, to the rehabilitation and upgrade of existing infrastructure. State and local agencies continually struggle with the question of how best to rebuild deteriorating roadways. Rehabilitating *flexible* pavements is especially important, because 94% of U.S. roadways are flexible (either surfaced or unpaved (FHWA, 1998). Achieving good performance at a reasonable cost is of paramount importance to the design engineer. The cost burden is

---

[1] Program Manager, Soil-Cement/Roller-Compacted Concrete Pavements, Portland Cement Association, P.O. Box 2615, Sugar Land, TX 77287-2615.

particularly acute for local road agencies. Counties, cities and towns are responsible for more than 75% of the centerline miles of roads in the U.S. These agencies are often significantly underfunded. They need a low cost, road rehabilitation strategy that requires minimal maintenance. Table 1 discusses potential solutions that can provide long-lasting improvement.

Table 1 - Characteristics of Flexible Pavement Rehabilitation Strategies

Solution	Advantages	Disadvantages
Thick Structural Overlay	• Provides new pavement structure • Quick construction • Only moderate traffic disruption	• Elevation change can present problems for existing curb & gutter and overhead clearances. • Large quantity of material must be transported in • Old base/subgrade may still need improvement. • High cost alternative
Removal and Replacement	• Provides new pavement structure • Failed base and subgrade are eliminated • Existing road profile/elevation can be maintained.	• Long construction cycle requiring detours and inconvenience to local residents/businesses • Increased traffic congestion due to detours, construction traffic • Rain or snow can significantly postpone completion • Large quantity of material must be transported in • Old materials must be dumped • Highest cost alternative
Recycling Surface Base and Subgrade and with Cement (Full-Depth Reclamation)	• Provides new pavement structure • Fast construction cycle • No detours • Minimal change in elevation, thus eliminating problems with curb/gutter, overhead clearances • Minimal material transported in or out • Conserves resources by recycling previously purchased materials • Local traffic returns quickly • Rain does not affect construction schedules significantly • Provides moisture- and frost-resistant base • Least cost alternative	• May require additional effort to correct subgrade problems • Some shrinkage cracks may reflect through bituminous surface

The Recycling Alternative

Recycling with cement is a process that involves pulverization of an existing bituminous surface along with the base and, sometimes, subgrade. A new thin bituminous surface course (asphalt or chip-seal) is then applied. This simple, quick and inexpensive procedure produces a high-capacity, durable pavement at low cost. It is sometimes referred to as "full-depth reclamation."

Recycling the existing surface and base of a flexible pavement with cement provides numerous advantages for the contractor, the owner/agency and the public. Because in-situ materials are re-used, very little material needs to be hauled off site and wasted. The road agency uses material it has already paid for, thus making best use of the public's money. Dwindling supplies of virgin aggregates are thus conserved for higher-end uses, because little new material needs to be hauled in. Cost savings resulting from avoided aggregate purchase, hauling of both new and waste material, and dumping fees can be significant.

The recycling operation itself is quick and therefore has minimal impact on the traveling public. No preliminary patching or rehabilitation is necessary. A section of road can be pulverized, blended with cement, compacted and cured all in a one-day operation. Local traffic can return to an unsurfaced, recycled base upon completion of compaction. Additionally, once a recycled base is fully compacted, it is resistant to rain, thus allowing construction to continue shortly after a shower or storm. An asphalt or chip seal surface course can be applied immediately, or at a later date.

Besides being quick and cost-effective, recycling with cement is versatile. On many older roads, pavement section materials can be highly variable, due to different road construction and maintenance techniques used over the years. Often, the engineer is not sure what types of materials will be encountered on the project (even sampling can leave information gaps). Cement has the ability to successfully stabilize and strengthen a wide spectrum of materials: Granular aggregates, pulverized asphalt, gravels, sands, silts and clays (ACI, 1990). Even when clayey, plastic subgrade material infiltrates and contaminates an old graded aggregate base course, cement will reduce the plasticity of the clay and allow it to be incorporated in the recycled stabilized base.

Cement is able to provide these benefits for both granular and fine-grained materials because of its chemical make-up. Cementitious reaction products— calcium-silicate-hydrate and calcium-aluminate-hydrate—form quickly when cement combines with water. These compounds act as a "glue" to bind materials together, reduce moisture flow channels, and improve gradation of fine-grained or poorly graded materials. Additionally, for active clay materials, cement produces calcium hydroxide, which causes a cation exchange in the clay and initiates flocculation of the clay particles. This dramatically reduces plasticity, moisture susceptibility and shrink/swell potential, while significantly improves bearing capacity (Prusinski and Bhattacharja, 1998). Additionally, the ability of cement to bind and confine the rehabilitated base materials will preclude future infiltration of clay into the base. Overall strength and performance of the stabilized base will thus be enhanced.

Finally, recycling with cement can appreciably improve the structural capacity and performance of the original roadway. A cement-stabilized base spreads traffic loads to the subgrade, thus reducing subgrade stresses and the potential for future failures, Fig. 1. Additionally, a cement-stabilized base maintains its structural integrity even in saturated conditions. An unbound aggregate base, on the other

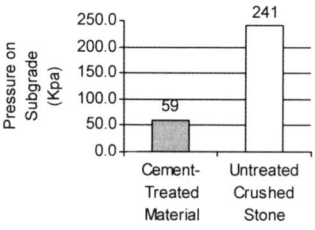

Fig. 1: Pressure at subgrade from a 550 Kpa (80 psi) surface load. Cement stabilization spreads loads to wider areas than unstabilized materials, thus significantly lowering subgrade stresses and reducing failures due to weak subgrades (PCA, 1962)

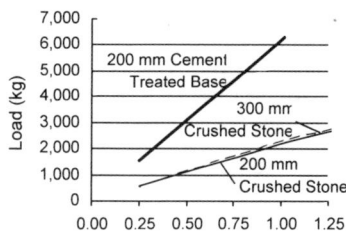

Fig. 2: Load-Deflection characteristics for cement-treated and crushed stone bases show higher stiffness of stabilized material, even with thicker crushed stone. (Nussbaum and Larsen 1965).

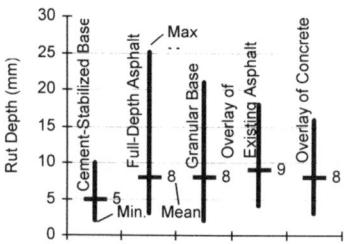

Fig. 3 - Rut depths of 10 to 15 year-old asphalt pavements with various base types, 10 to 15 years old. Cement stabilization produces the lowest mean, maximum and minimum rut depths (Brent Rauhut Engr, 1996).

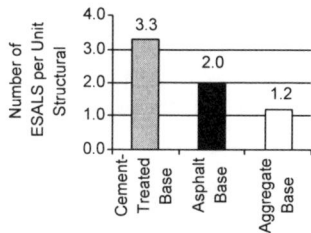

Fig. 4: Mississippi pavement data indicates that cement treatment provides higher levels of trafficking prior to rehabilitation (George, 1999).

hand, is highly moisture susceptible and loses much of its capacity during times of saturation.

The stabilized base is much stiffer than an equivalent unbound aggregate base, thus subjecting the surface to less deflection, Fig. 2, which lowers fatigue stress in the asphalt and resultant surface cracking. Finally, rutting caused by shifting and consolidation of base materials is virtually eliminated in cement-treated bases. Total rutting is much less, as it is confined to the bituminous surface course, Fig. 3. With recycling, if traffic loads have increased, the pavement section capacity can be improved simply by pulverizing deeper into the subgrade and adding the appropriate amount of cement to stabilize the combined materials.

Cement stabilized pavements will shrink slightly (similar to concrete) due to cement hydration, moisture loss and strength gain. This often manifests itself as transverse cracks with a 10 ft to 40 ft spacing. Most cracks are narrower than 2 mm (1/8" wide). Some cracks may reflect over time through the bituminous surface. Fine cracks usually are not a performance concern and can be ignored. Crack minimization strategies are discussed in the section on *Quality Control*.

Cement treatment of bases provides excellent long-term durability. A recent study (George, 1999), comparing the performance life of in-service aggregate, asphalt-treated and cement-treated bases in the State of Mississippi found that bases stabilized with cement can sustain 65% and 175% higher load repetitions (per unit structural number) than asphalt and unbound aggregate base, respectively, Fig. 4.

Mix Proportions

Determining the amount of cement to properly stabilize recycled/in-situ materials generally consists of two steps: 1) Finding the moisture-density relationship and 2) Determining cement content to provide strength and durability. To calculate the optimum moisture and maximum dry density, a representative sample of materials is pulverized to the anticipated gradation (normally 100% passing a 2" sieve and 55% passing a #4 sieve, exclusive of gravel). An initial cement content is estimated and added (PCA 1992 can provide guidance). A moisture-density curve is then developed using the standard Proctor procedure (ASTM D558).

Next, a series of specimens is molded using three or more cement contents at, above and below the initial estimate. One of the following methods can then be used to determine adequate cement content (for detailed procedures, see *Soil-Cement Laboratory Handcbook*, PCA, 1992):

- **Durability Tests (ASTM D559/D560):** These tests subject samples of stabilized materials to a 12-cycle wetting/drying and freezing/thawing regimen. After each cycle, the sample is wire brushed in a specified manner. After the final cycle, the specimen is dried completely, and the amount of material loss is measured. Cement content is based on the minimum required to meet a recommended loss criterion for a specific AASHTO soil type. Durability tests normally take about 28 days.
- **Unconfined Compressive Strength (UCS) Tests (ASTM D1633):** Many state and local transportation agencies have opted to use strength tests in lieu of the durability tests (often based on correlations among durability tests, UCS tests and field performance). Seven-day strength criteria generally fall between 1.4 and 5.5 Mpa (200 and 800 psi) (ACI, 1990). Recent research has indicated that cement contents which provide lower to medium levels of strength—1.7 Mpa to 3.4 Mpa (250 to 500 psi)—provide good performance, without too much strength to make the material overly rigid and susceptible to wide cracks and faulting (Saarenketo, and Scullion 1995).

- **Rapid Test Procedure.** For small jobs or emergency construction, a qualitative test method—often referred to as the "pick and click" test—was developed by PCA. This test takes no more than two days and can be accomplished without any laboratory equipment.

The Texas Transportation Institute of Texas A&M University is currently developing new tests which base cement content on a combination of moisture susceptibility, strength and shrinkage of soil-cement specimens (Saarenketo, and Scullion 1995).

Thickness Determination

For thickness design, AASHTO design methods for flexible pavement design can be used. A layer coefficient is assigned to cement-stabilized materials by the specifying agency. This coefficient is multiplied by base depth to yield that layer's contribution to a "structural number." The required structural number is based on traffic and service conditions. (AASHTO 1993) Examples of soil-cement layer coefficients for several state DOT's are provided in ACI, 1990 which range from 0.12 to 0.28, with most being in the range of 0.17 to 0.23.

The PCA has developed a thickness design method for soil-cement (PCA, 1970), which can be applied to recycled pavements. This procedure allows the designer to adopt a loading spectrum based on traffic counts and axle loads, and a design thickness which takes into account subgrade strength, soil type and fatigue consumption.

Because the base is the principal load-bearing structure for a recycled pavement, recommended bituminous surface thickness is relatively thin—only enough to provide surface durability under expected traffic, and protection from moisture. Table 2 provides typical base and surface values for various types of traffic.

Table 2 - Typical Base and Surface Requirements For Roads Recycled with Cement

Road Function	Typical Thickness	Recommended Surface
Residential	125 mm (5 in)	20 – 40 mm (0.75 – 1.5 in)
Secondary	200 mm (8 in)	40 to 65 mm (1.5 – 2.5 in)
Highway	250 mm (10 in)	50 to 75 mm (2 to 3 in)

Note: Surface can be combination of asphalt and/or bituminous surface treatment. (PCA 1970)

The Recycling Process

Recycling a failed flexible pavement is quick, straightforward and economical. Methods follow similar procedures employed for construction and quality control of traditional mixed-in-place soil-cement (PCA 1995). Recycling consists of the following steps:

1. The old asphalt surface is ripped down to base level with the ripper teeth of a grader. If the asphalt layer is deep, some of the top surface may need to be milled off first.

2. The asphalt surface and underlying base materials are broken down through the use of a single shaft rotary mixer, Fig. 5, often referred to as a "pulver-mixer." These machines have shafts or drums rotating at high speed, with numerous teeth which break down and pulverize the pavement materials, Fig. 6. Several passes may be required to sufficiently pulverize the material so that 100% passes a 2 in. screen and at least 55% passes the #4 sieve. Some mixers are powerful enough to break down the surface without the preliminary ripping described in step 1. After pulverization, the material is shaped to proper crown and grade.

Fig. 5: Single transverse-shaft rotary mixer capable of pulverizing asphalt/base and mixing cement in a one-pass operation.

Fig. 6: Mixer teeth provide thorough ripping and pulverization.

3. A measured amount of cement is spread on top of the pulverized and shaped material. Cement can be spread dry, 1) through the spreader bar attached to a pneumatic tanker, 2) through a "Flynn" type gravity spreader, fed by the pneumatic tanker and metered out onto the surface below, or 3) through a spreader truck with metered gravity feed. Bags of cement can be used if bulk cement is not readily available. Cement can also be applied as a slurry to reduce dusting.

4. The cement is then blended into the pre-pulverized asphalt/base/subgrade to the appropriate depth with a pulverizer. Water, up to the material's optimum moisture content, can be added at this time directly through the mixing chamber. Alternatively, water can be applied externally with a water truck, subsequent to the dry cement mixing pass, and then blended in with the pulverizer.

5. The cement-treated material is then compacted to maximum density using equipment that can include vibratory steel drum and pad-foot rollers, and pneumatic tire rollers.

6. After compaction, the surface is shaped to final crown and grade. The surface is then re-rolled.

7. Proper curing is essential to maintain adequate strength and minimize shrinkage cracking. Therefore, after finish grading, water is applied periodically to the surface to ensure that the material does not dry out.

8. A prime coat of bituminous material (such as SS1 or MC30) usually is spread at the rate of 0.15 to 0.30 gallons/square yard to seal the surface. Normally, this is done at the end of the construction day so that water curing need not continue. If local or construction traffic is to be applied, the prime coat should be sanded to prevent pick-up on the tires of the traffic.

9. A thin asphalt or chip-seal surface can be applied any time thereafter. If scheduling permits, a curing period from seven to 28 days is recommended for an unsurfaced cement-treated base. However, if the road cannot be closed, the surface can be applied the next day and traffic can proceed as long as the base has reached at least 200 psi.

There are alternative construction processes to the one described above. Sometimes the entire process—cement application, pulverization, moisture addition and compaction—is performed in a one pass operation (see figures 5 and 6). In another method, stockpiled asphalt millings are mixed with water and cement in a central plant pug mill and trucked to the jobsite. In a third process, a paving train mills the asphalt/base, crushes/screens the material, adds cement and water in a pug mill, and lays down the mixture.

Quality Control

To ensure that the desired results are achieved, proper specification and QA/QC procedures must be followed. These procedures, summarized below, are fully discussed in several references (PCA 1980, 1995(1), 1995(2)):

1. Prior to recycling, the engineer should analyze the causes of pavement distress and determine if recycling is an appropriate solution. In most cases, upgrading the base through recycling will solve the principal problems such as rutting, base deterioration and surface fatigue cracking. However, some problems, like subgrade failure, must be remedied prior to recycling, or another rehabilitation strategy employed.

2. If an asphalt mat retains most of its original viscosity and flexibility, it should be removed completely, because proper pulverization and subsequent bonding with cement will be difficult. Brittle, "dead" asphalt or chip-seal surfaces are the best candidates for recycling. Some pulverizers, however, are powerful enough to sufficiently break up even recently placed asphalt mats.

3. Gradation of the pulverized base should meet the requirement of 100% passing the 2 in. sieve and 55% passing the #4 sieve.

4. Cement spread during construction should be checked to ensure the specified amount is placed atop the pulverized base.

5. Moisture content and density should be checked after compaction (normally with a nuclear gauge) to ensure that moisture content is within 2 points of optimum, and that density is a minimum of 96% of maximum dry density, based on a standard Proctor.

6. Time delay between mixing and compaction must be held to 3 hours or less. If the material is to remain uncompacted for 1 hour or more, it should be re-mixed every 30 minutes.
7. Curing is critical. The surface, once compacted, should never be allowed to completely dry out. A bituminous cure coat ends the necessity for periodic water applications.
8. Fine shrinkage cracks will occur in the stabilized base, and may reflect through the surface. This is normally a cosmetic manifestation, and not a performance concern. Cracks 1/8" or more should be sealed using standard methods. Cracking and reflectivity can be minimized through (Kuhlman 1994, George 1999):
   - Following proper quality procedures as described above, especially with regard to mixture homogeneity, achieving proper moisture and density, and curing.
   - Keeping moisture content at or slightly below optimum moisture content (+0% to −2% optimum).
   - Providing a 7 to 28 day curing period to allow initial cracking to occur prior to surfacing.
   - Using a chip-seal surface or interlayer (such as a geotechnical fabric or unstabilized aggregate) between the base and asphalt surface, thus providing an elastic layer to reduce reflectivity.
   - "Precracking" through 1) early surfacing/traffic application, 2) re-application of compaction equipment after 1-2 days, or 3) inducing weakened planes in the material by grooving mixed but uncompacted material at about 3 m (10 ft) intervals and injecting asphalt emulsion.

Table 3 - Cost Estimates for Recycling vs. Removal/Replacement

Cost Item	Recycling with Cement cost/sq m (cost/sq yd)	Removal and Replacement with New Base cost/sq m (cost/sq yd)
Cement	$0.87 to $2.65 ($0.72 to $2.21)	--
Aggregate	--	$1.82 to $5.46 ($1.52 to $4.57)
Processing	$1.20 to $2.99 ($1.00 to $2.50)	$0.90 to $2.39 ($0.75 to $2.00)
Haul	--	$0.91 to $2.34 ($0.76 to $1.96)
Landfill or Stockpile	--	$0.73 to $23.81 ($0.61 to $19.90)
Total	**$2.07 to $5.64 ($1.72 to $4.71)**	**$4.35 to $34.00 ($3.64 to $28.43)**

Note: Costs are based on estimates and procedures from ENR, 1999, BioCycle, 1998, and Peurifoy & Oberlender, 1989. Assumed base thicknesses for equivalent load capacity are 150 mm (6 in) thick base course for recycled material and 200 mm (8 in) thick graded aggregate base course. Costs do not include subgrade preparation nor surface course.

An Economical Alternative

The cost of recycling is generally less—often significantly less—than other methods of road reconstruction. Cost ranges for material and processing costs on ce-ment-recycled pavements versus removal/ replacement with new material are as shown in Table 3, above. Note especially the wide range of costs possible for stock-piling or landfilling construction debris. In some jurisdictions, these materials can no longer be landfilled, creating a significant problem with growing "stockpiles." Con-gested urban areas are especially prone to the high costs of landfilling/stockpiling.

U.S. Recycling Case Studies

Texas DOT's Bryan District.   The Texas Department of Transportation's Bryan District is a mostly rural district with numerous structurally deficient Farm-to-Market roads. Agricultural and oil field traffic have increased the loads on these roads (mostly thin bituminous surface on unstabilized bases), and caused many maintenance problems. The district is distant from inexpensive sources of quality aggregate, so removal and replacement of material is costly. The district's chosen alternative is a systematic rehabilitation of FM roads through stabilization. The objective of stabilization is to correct defects and increase capacity. A typical recycled section consists of 10" of stabilized recycled base with a two-course chip seal surface.

The old asphalt surface is completely milled and removed prior to base recycling. For higher-capacity roads, an additional layer of 4" crushed rock is added between the stabilized material and the surface course. The district chose to not follow standard Texas DOT specifications. These specs call for up to 5.2 Mpa (750 psi) strength, which is now thought to be too strong, resulting in rigid behavior and excess shrinkage. Alternatives used by some districts include 1.4 MPa (200 psi) strength or achieving three times the unstabilized compressive strength of the materials. Ultimately, cement contents on the low side, around 4%, are chosen. The roads are 2-lane, and usually provide the only access to a particular area, thus they cannot be closed down at all. At the end of the construction day, the roads are opened to full traffic.

The Texas Transportation Institute evaluated performance of the pavements over time. 25 projects were examined, all being four years old or less. The study concluded that the base recycling strategy works well:
- A stiff base is produced which generally exhibited little distress, and virtually no shrinkage cracking.
- Stabilization provides waterproofing and confinement of the base material
- Too-high strength may produce problems because of low strain capacity of the material. High strength, stabilized materials are brittle and can crack in a rigid manner. The worst situation occurs with clay subgrades (PI > 30) with trees less than 30' to the pavement edge. Moderate to low strength material, with higher strain capacity, is recommended.

Dallas County, TX. Dallas County is a highly urbanized area with significant traffic congestion. The county is able to take advantage of recycled asphalt pavement (RAP) millings "donated" by the TxDOT (County Progress, 1998). The county hauls the millings to a cement-treated base plant where the material is processed in a pug mill, and blended with measured amounts of cement. This fast and efficient processing method produces a high-quality base material. Because of continuous monitoring of material additions (RAP, cement and water), material proportions are easier to control and are generally more consistent than mixed-in-place materials. Also, field processing time is decreased by up to 80%, because mixing is accomplished off-site. The county has found that the asphalt millings are graded finely enough so that in most cases, additional fine granular material does not need to be added. Occa-

sionally, sand is added if a seam of RAP is too coarse; the flexibility of the feeder operations makes this an easy task. Oversized material is screened out at the plant. Roads placed in this manner have performed very well according to the Dallas County Maintenance Department, and have resulted in reduced expenses and road closure time, Fig. 7.

Fig. 7: RAP, blended with cement offsite in a pug mill, is placed and compacted in Dallas county

Waco, TX. Many of Waco, TX's residential streets, which consist of granular base with a thin bituminous surface, have reached the end of their service lives. The city has recently chosen in-place recycling with cement as a rehabilitation

strategy, because of the cost savings and durability it offered. Instead of dry application, Waco chose to specify cement slurry to minimize dusting in its residential neighborhoods.

Cement (from a pneumatic tanker) and water are fed simultaneously into a

Fig. 8: Cement slurry applied to pulverized asphalt and base.

mixing pump which thoroughly blends the materials. The slurry is pumped into a water truck. A 50/50 blend, by weight, of cement and water is used. The slurry is spread on the pre-pulverized base/ surface (which sometimes includes subgrade material when the existing base thickness is not adequate) within ½ hour of mixing with water, Fig 8. Pulver-mixing commences immediately, followed by compaction and curing. Local traffic can return to the pavement as soon as compaction is complete. A thin asphalt surface is applied to complete the construction.

Spokane County, WA. Much of the rural road system in Spokane County, WA evolved from buggy track. Most were 6.5 m (21 ft) wide, and a portion had been surfaced with bituminous surface treatments (BST). Over the past three years, Spokanne developed a pavement management program which gathered data and priortized road repair/reconstruction projects. Most roads are maintained by blade patching and chip sealing. However, roads maintained in this manner have load restrictions imposed each Spring because of deficient structure. For many roads this is acceptable, as they are very low volume. However, some roads support business operations, which would not be able to function with the load restrictions imposed. For these higher duty roads, the county chose to recycle the materials in place with cement. This method would provide higher load carrying capacity year-round, and would eliminate Spring load restrictions. The recycling process offered the additional advantage of not significantly changing the profile elevation, thus eliminating the need for expensive fills on the sides to maintain the same road width. A BST surface was planned since asphalt would have been cost-prohibitive.

The county does not have its own pulverizers, so it developed a "joint venture" arrangement with a stabilization contractor. The contractor was responsible for pulverizing the road and blending cement to a 200 mm (8 in) depth. 4% cement was chosen, as this would provide a moderate strength of 3.0 MPa (430 psi), yet would allow shrinkage cracks to form relatively closely spaced, but remain narrow and tight. The county was responsible for water application, blading, compaction and finishing. Two CMI 650 pulverizers were utilized to recycle the road in 300 m (1,000 ft) increments. This length was chosen because it allowed mixing, compaction and finishing to occur in a timely manner, before the cement began to set. The material was compacted to 95% standard proctor density, was shot with a curing seal of CSS-1 (50% water dilution). Car traffic was allowed on the road almost immediately, but trucks were restricted for two weeks. Additionally, about 75 mm of WaDOT "top course" aggregate was placed on the base after about three weeks, because some of the recycled underlying material was oversized (greater than 50 mm [2 in]) and would not provide a smooth final surface when finish graded. The final stage was to apply a two-course BST chip-seal. A third course is planned for 2000.

Cost for the 7 km (4.4 mile), 7.3 m(24 ft) wide road rehabilitation project was only $91,000 per mile. This included $34,000 for the contractor to pulverize, furnish cement and mix, $22,000 for the County's portion of the base work (including top course gravel) and $35,000 for the BST/chip seals, including the third course projected in 2000. In comparison, the county estimated that reconstruction using a gravel section would have cost $135,000 per mile, nearly 50% more than recycling. Because of this project's success, Spokane County is planning more cement recycling in 2000 and beyond.

Westminster, CA. The City of Westminster, CA recently chose to use cement for recycling of alligator-cracked streets. The city is in an area with poor subgrade and a high water table. It needed a solution that minimized inconvenience for local residents, did not impact curb and gutter elevations, and provided a durable pavement

structure even under moist conditions (Cartwright, 1999). For a severely deteriorated pavement like that pictured in Fig. 9 an overlay was not deemed a durable fix—a full base reconstruction was necessary. However, removal and replacement with new material would cause significant construction traffic, and would necessitate closing of streets for long periods of time. Recycling with cement was an attractive choice since it minimized traffic and excavation; it also increased the strength of the roadway without adding additional section thickness, and did not alter the existing road profile.

Fig 9: Failed asphalt pavement in Westminister. CA.

Fig. 10: Cement is spread prior to mixing into the pulverized base.

In this 1998 project, the existing asphalt varied from 75 to 125 mm (3 to 5 in), and base varied from 100 to 150 mm (4 to 6 in) in thickness. The asphalt surface and granular base, along with some underlying subgrade, were pre-pulverized and blended to a depth of 300 mm (12 in) prior to stabilization. Only about 65 mm (2.5 in) of material had to be hauled away after re-grading to bring the base to the proper elevation. Cement (6% by weight, Fig. 10) was then blended along with water into the material, compacted and water cured from five to seven days. Residents could drive on the completed base and easily pull into their driveways since the drop-off was only 65 mm (2.5 in) prior to surfacing. Even during construction, the road never had to be completely closed. The city chose to use a geotextile fabric sandwiched between two thin layers of asphalt to preclude reflection cracking in the surface.

Florida DOT. The Florida Department of Transportation uses a native rock, termed "limerock" as a base material for much of its road construction (Turner, 1999). The rock is chemically similar to limestone, but is internally porous, somewhat soft (Los Angeles abrasion from 30% to 50%) and absorptive (3% to 6%). When dry, limerock is strong and durable. However, limerock has a high moisture affinity and, when saturated, becomes soft and deforms under loading.

Rehab of older roads is often accomplished by milling the old asphalt to within 12 mm (0.5 in) of the limerock base, then resurfacing. The old base is left undisturbed, with a "protective" cover (the old, oxidized asphalt) during construction, in order to maintain the integrity and strength of the underlying limerock (which gains strength with age). More than 14 years ago, one project being rehabilitated in the above manner (SR-349) in District 2 became subjected to heavy rain, resulting in breakage of the 12 mm (0.5 in) asphalt cover, and severe deterioration of the base.

District engineers decided that to restore the integrity of the base, they would pulverize the existing base and new asphalt cap together, with a small amount of cement. This would reduce moisture susceptibility, and promote strength gain, but would not be enough cement to allow reflective cracking. Within two days, the problem section had been mixed with cement, rolled and graded. This section has now completed over 14 years of service with no deterioration.

Since the original section was placed, use of reclaimed base has expanded, and has been used in at least 13 District 2 projects. For pavement design, a layer coefficient for cement-recycled material had to be developed. The Florida DOT back-calculated the layer coefficient by using actual vehicle traffic counts and loads—coupled with the success of early projects—to answer the question, "What value is required to cause the structural number to predict success under the loading?" The back-calculation yielded a layer coefficient of 0.33 per inch and several subsequent projects have been built using a coefficient of 0.30. These projects continue to perform well. In fact all but one project has provided excellent performance. (The sole exception occurred when material was mixed during a severe freeze with (possibly) high moisture content.)

Conclusion

Recycling failed flexible pavements with cement is highly economical, and ultimately produces an improved base. Additionally, recycling is "environmentally friendly," virgin aggregates are conserved and landfilling of old material is avoided. Cement treatment improves strength, repels water, and provides increased stiffness, which prolongs pavement life. Many recycling projects have been successfully constructed throughout the U.S. at the state, county and municipal levels. Recycling is an alternative that deserves serious consideration for any flexible pavement rehabilitation project.

References

American Association of State Highway and Transportation Officials, "Guide for Design of Pavement Structures" (1993)

American Concrete Institute, "State-of-the-Art Report On Soil Cement," ACI 230.1R-90 (1990)

California Asphalt, "City of Westminster's Reconstruction Project Saves Time and Money, And Pays Huge Dividenes," (March/April 1999).

CMI Corporation, "A Practical Guide to Soil Stabilization & Road Reclamation Techniques" (September 1998)

County Progress Magazine, "Rehabing Dallas County Roads," (September 1998).

Engineering News Record, "Construction Economics" Cost Index Table, p. 43, (September 6, 1999).

Federal Highway Administration, "Highway Statistics," Table HM-12, (December 1998).

George, K. P. "Minimizing Cracking in Cement-Treated Materials for Improved Performance," Portland Cement Association (February 2000).

Kuhlman, R. H., "Cracking in Soil-Cement—Cause, Effect, Control," *Concrete International* (August 1994)

Nussbaum, P. and Larsen, T., "Load-Deflection Characteristics of Soil-Cement Pavements," Highway Research Record No. 86, Highway Research Board (1965).

Peurifoy, R.L. and Oberlender, G.D., *Estimating Construction Costs Fourth Ed.*, McGraw-Hill, Boston (1989).

Portland Cement Association, "Recycling Failed Flexible Pavements with Cement," Report IS197 (1993).

Portland Cement Association, "Soil-Cement Construction Handbook," Report EB003 (1995).

Portland Cement Association, "Soil-Cement Inspector's Manual," Report PA050 (1980).

Portland Cement Association, "Soil-Cement Laboratory Handbook," Report EB052 (1992).

Portland Cement Association, "Suggested Specifications for Soil-Cement Base Course," Report IS008 (1995).

Portland Cement Association, "Thickness Design for Soil-Cement Pavements," Report EB068 (1970).

Prusinski J.R. and Bhattacharja, S., "Effectiveness of Portland Cement and Lime in Stabilizing Clay Soils," Transportation Research Record 1652, Volume 2, Transportation Research Board (1999).

Saarenketo, T. and Scullion, T., "Using Electrical Properties to Classify the Strength Properties of Base Course Aggregates," Report 1341-2, Texas Transportation Institute, College Station, TX (November 1995).

Turner, D.R., "Reclaimed Base," Florida Department of Transportation District II internal report (1999).

Relocation of The Cape Hatteras Light Station –
Move Route Design and Construction

J. Allan Tice[1] and Randy A. Knott[2]

Abstract

Relocation of the Cape Hatteras Lighthouse and associated structures across a coastal dune system was planned by the National Park Service to protect the structures from continuing coastal erosion. The lighthouse is a masonry structure about 61m tall and weighing about 4,400 metric tons. Preparation of the loose sands along the move route required a combination of proofrolling, in-place densification and placement of compacted stone layers to control settlement during the move. A geotechnical site characterization program consisting of conventional exploratory borings supplemented by dilatometer testing and a large scale area load test was performed to determine the necessary ground improvement. Elastic analyses were performed to estimate settlements under the planned move system. Measurements of the deflections of the steel mats during the move were obtained and compared to predicted deflections.

Introduction

The Cape Hatteras Lighthouse, located on the Outer Banks of North Carolina, is a national historic monument managed by the National Park Service. It is also designated a National Civil Engineering Historic Landmark by ASCE. The masonry lighthouse was constructed in 1870, and, at a height of 61m, is the tallest lighthouse in the United States and the tallest masonry lighthouse in the world. Coastal erosion over many years caused a threat of eventual undermining of the lighthouse on its existing shallow stone and timber mat foundation. The National Park Service, after many years of study, determined that the lighthouse should be moved 884 meters to the southwest for long-term safety.

---

[1] Member, ASCE, Geotechnical Consultant, Law Engineering and Environmental Services, Inc., Raleigh, North Carolina 27604
[2] Fellow, ASCE, Chief Engineer, Law Engineering and Environmental Services, Inc., Kennesaw, Georgia 30144

The relocation contractor used steel shoring and beams to temporarily support the 4,400 metric ton lighthouse. Under the steel beams, seven main beams with hydraulic jacks and rollers transferred the load to seven steel track beams, which were in turn underlain by steel matting placed on the improved ground. The hydraulic jacks and a three-zone unified pressure control system could accommodate some variations in ground surface deformation without causing the lighthouse above to deviate from a level plane.

The proposed move route crossed existing dune and former beach sands. The upper few feet were loose with relatively dense sands below. The ground-water table was within 1.5 to 2 m of the surface. The primary geotechnical challenge of the project was how to prepare the sands to provide relatively uniform support with reasonably reliable predicted settlements while maintaining costs within a lump sum project budget. In addition, procedures to allow the transition from the existing foundation (a timber mat) to the move route, and then from the move route on to the new foundation (a reinforced concrete mat) were needed that would keep deflections below limits established by the structural and relocation engineers.

The National Park Service executed a design-build, lump sum fee contract with a team led by International Chimney Corporation of Buffalo, NY to design and perform the move. Law Engineering and Environmental Services, Inc. (LAW) provided geotechnical, environmental and overall quality control/assurance services for the project.

Geologic Setting

The lighthouse site is located in the Coastal Plain Physiographic Province of North Carolina in a region known as the Outer Banks (Fig. 1). The Coastal Plain

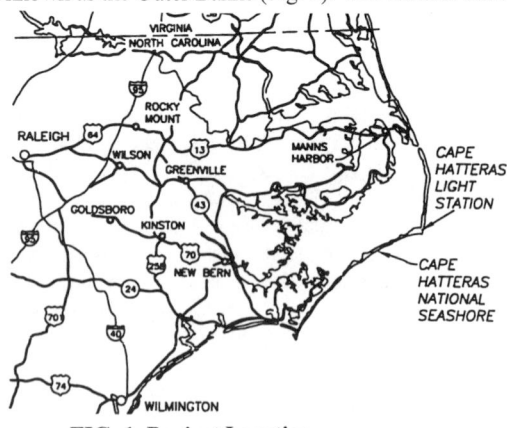

FIG. 1 Project Location

consists mainly of marine sediments which were deposited during successive periods of fluctuating sea level and moving shoreline. Since construction in 1870, the shoreline at the lighthouse location has moved west about 450 m (Fig. 2).

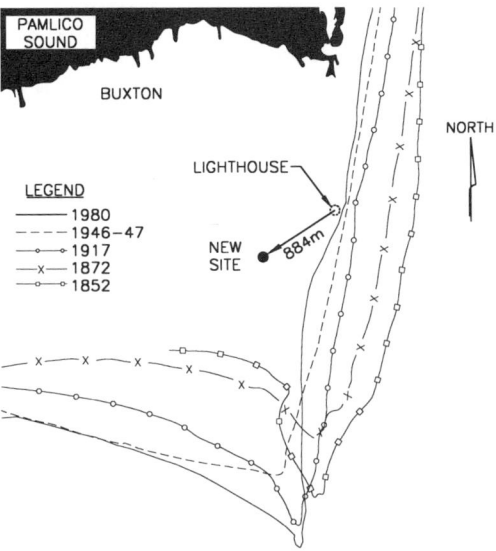

FIG. 2 Shoreline Regression at Hatteras Island. Adapted from Figure 7 in 1988 National Research Council Report "Saving Cape Hatteras Lighthouse from the Sea".

The geologic formations dip slightly seaward and several are exposed at the surface in bands paralleling the coast. Many formations exist only as fragmental erosional remnants sandwiched between more continuous strata above and below. The soils in this province include sands, silts, and clays with irregular deposits of shells and organics.

The depositional pattern in the Outer Banks geologic setting is the result of a barrier island – lagoon system in which the barrier islands separate quiet Pamlico Sound waters from active Atlantic Ocean waters. The barrier island strata are predominantly sands, and the lagoon strata are typically clays, silts, and fine sands. Most near-surface soils in both settings are normally consolidated, having never been subjected to past overburden stresses greater than the present stresses. Zones of soft, compressible silt near the surface are common on the lagoon side of barrier island dunes. Fig. 3 shows a schematic geologic section.

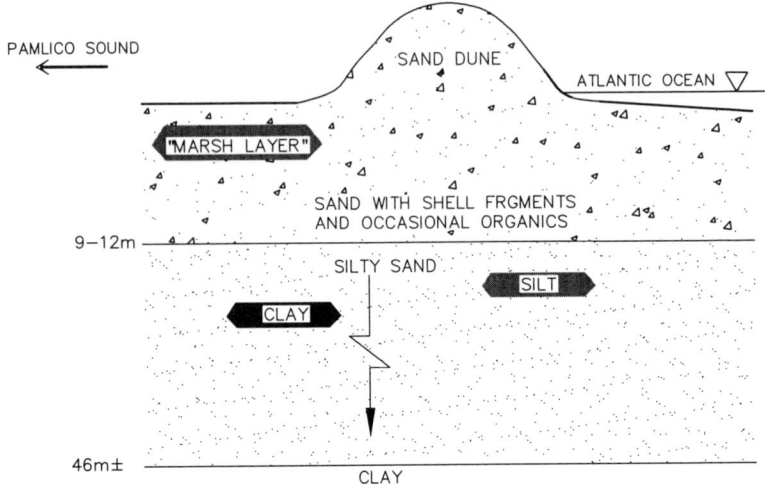

FIG 3. Schematic Geologic Section

Geotechnical Data Collection

Soil descriptions, profiles and standard penetration test (SPT) data were available for most of the move route from earlier work by the U.S. Army Corps of Engineers. LAW supplemented these data with 10 additional borings with SPT to depths of 4.5 to 23 m. Borings were made using rotary wash drilling with bentonite-weighted drilling fluid. Standard penetration testing was conducted at 0.75 m intervals to 6 m, then at 1.5 m intervals to maximum depths of 23 m.

Because the soils were mainly relatively clean sands, the settlement performance was expected to be related to the elastic properties of the sand. In-situ soundings using the Marchetti flat-plate dilatometer test (DMT) was performed at 21 locations spaced along the move corridor to depths of 2.5 to 5 m during the design phase of the project to obtain elastic modulus values for use in analysis. The tests were performed at 0.3 m intervals.

Later, during construction, an additional 85 shallow depth (less than 2 m) DMT soundings were made to aid in locating local areas of very loose sands with organics, and to provide data for settlement evaluation of these areas. The additional sounding locations were positioned on each side of the lighthouse move path and spaced such that, at any one position of the lighthouse footprint, at least two soundings would be made. The test interval for these additional soundings was 0.15 m.

After analysis using the design-phase DMT data, a large-scale field load test was conducted to check the analytical results and to provide additional information on the settlement potential of the sands. A location was selected where boring and DMT data indicated a possible thin, loose layer at a relatively shallow depth. LAW constructed a 4.6 m tall, conical pile of sand in three 1.5 m+/- increments. The diameter of the pile base was approximately 18 m and side slope angle measurements ranged from $27°$ to $36°$. The pile was made relatively large in order to affect the soils to as great a depth as was feasible, given the site constraints. Prior to constructing the pile, an area measuring roughly 36.5 m by 21 m was cleared and grubbed, and the pile footprint was densified with a relatively small (1m diameter drum) vibratory roller making four passes in two perpendicular directions to simulate the expected future surface preparation.

Settlements of the ground surface under the pile were measured using differential leveling techniques and monitoring of two settlement plates placed at the original ground surface near the center of the pile. The settlement plates consisted of 1.2 m by 1.2 m, 1.9 cm thick plywood sections with a 1.9 cm diameter threaded flange and steel pipe attached to the top of the plywood. Direct measurements of the elevations at the tops of the pipes were made prior to construction of the pile, after 1.5 m of material was placed, and after completion of the pile. Fig. 4 shows the general arrangement of the load test and the settlement devices.

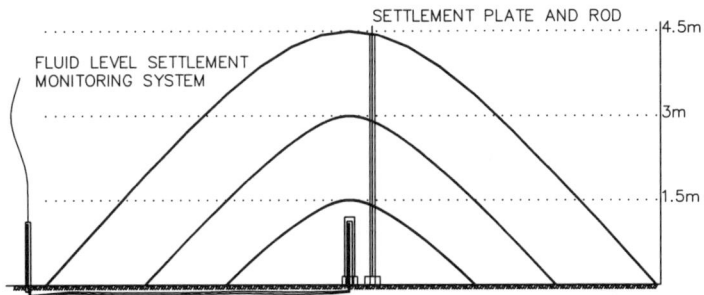

FIG. 4 Large Sand Pile Load Test Schematic

A water level device was integrated with one of the settlement plates to allow continued settlement measurements as the pile was built. A section of 1.3 cm vinyl tubing was fed vertically through the threaded pipe to an outlet spout located about 1.2 m above the top of the plywood plate. A 5 cm PVC casing was placed over the threaded pipe in order to provide a space for the spout to drain. The opposite end of the vinyl tubing was fed horizontally through PVC casing and placed in a shallow trench leading from the center of the pile to a measuring stand outside the limits of the pile. The tube was mounted vertically to a measuring device on the stand and was filled with water.

Measurements of the water level required slowly adding water to the tube until the water level reached equilibrium. A reading was taken prior to constructing the pile in order to establish a baseline (no settlement). Periodically during construction of the pile, additional measurements were made. As the soils beneath the pile settled, the outlet spout near the center of the pile dropped, and water poured from the tube; the water level on the measuring stand dropped by a corresponding amount.

A laboratory testing program was conducted to aid in soil classification, to determine soil index properties, and to provide data for development of settlement estimates and recommendations for bearing capacity, compaction and ground-water control. Laboratory testing of soil samples included the following:

- Grain size distribution on 24 samples
- Natural moisture content on 7 samples
- Atterberg limits (soil plasticity) on 4 samples
- Compaction (moisture-density relationship) on 2 samples

Soil Conditions

The subsurface explorations encountered generally consistent and uniform subsurface conditions across the site. The conditions were typical of the Outer Banks geologic setting. Loose (Standard Penetration Resistances, or N-values, ranging from 4 to 10) fine sand with a Unified Soil Classification System (USCS) designation of SP are present in the upper 1 to 3 m, with most loose sand present at depths of 2.5 m or less. Root fibers and decayed organics are common in the top 30 cm. Figure 5 illustrates the general subsurface soil profile.

Several borings encountered a thin layer of slightly silty, fine sand (SP-SM) with some organics. The depth to the top of this layer ranged from about 0.5 to about 2 m. The thickness of the layer appeared to be about 15 to 30 cm. Auger borings and additional DMT soundings as mentioned earlier were performed during the construction phase to further evaluate this layer. The additional testing found the layer was commonly present in two areas of the move route with generally a sandy texture and some organic material. The thickness rarely exceeded 15 cm, and the layer was generally present at a depth of about 1.2 m below the subgrade of the move route.

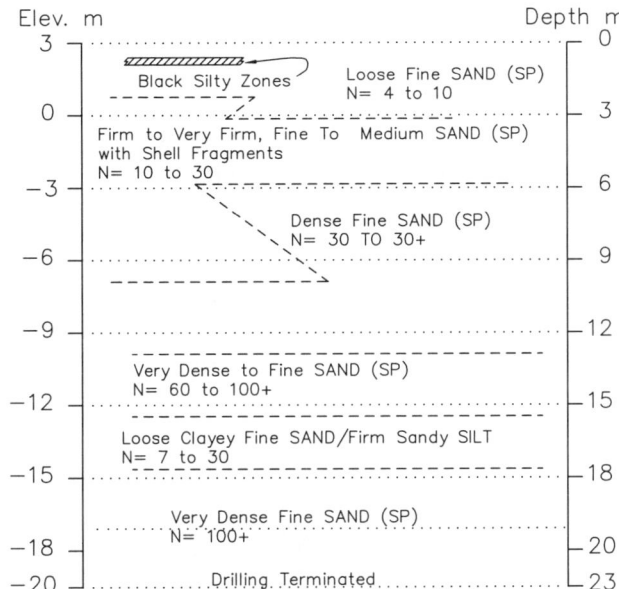

FIG. 5 General Subsurface Section for Move Route

The upper loose sands are underlain by firm to very dense (N-values ranging from 11 to over 100), fine and fine to medium sand (SP) with occasional shell fragments and fine gravel. The density of these underlying sands generally increases with depth, and very dense (100+ blow) material is common at depths below about 13.7 m. A layer of loose to firm silty, clayey, fine sand (SM/SC) was encountered between depths of about 16 to 17.5 m in some borings. Very dense sand is present below this layer, extending to the maximum depths explored.

Ground Water Conditions

Published geologic and water resource information suggested the Cape Hatteras area has a mounded fresh water lens atop salt water, with the water table at a relatively shallow depth. Water level measurements in temporary casings several days after boring completion indicated ground-water levels between 0.6 and 1.2 m below the ground surface. The ground-water measurements generally corresponded to depths at which soil sample moisture increased from moist to saturated. Comparing ground-water level measurements to the existing ground surface, ground water was indicated to be present between elevation 0.9 and 1.8 m, msl. Ground-water levels were measured at low and high tides to check for tidal influence. The measurements indicated ground-water fluctuations of only about 2.5 cm between low and high tides.

Soil Deformation Characteristics

The data from the SPT and the DMT showed loose to firm sands extending to depths of about 2.5 m to 3.5 m. The DMT provides a direct measurement of a soil's deformation (or strain) response to applied stress. The DMT modulus (M-value) represents a constrained modulus, and this value was used in calculating soil elastic deformation. The M-values were typically 10 to 30 Mp in the top 0.3 to 0.6 m, then increased to values generally between 50 and 100 Mp below 0.6 m. In the thin loose zones described earlier, the M-values ranged from about 2 to 10 Mp. Typically, the M-values immediately above and below the loose layer were greater than 40 Mp. Figure 6 shows the range of M-values in the upper 2.5 m, and Figure 7 shows typical M-values recorded in one of the shallow DMT soundings at a loose zone area.

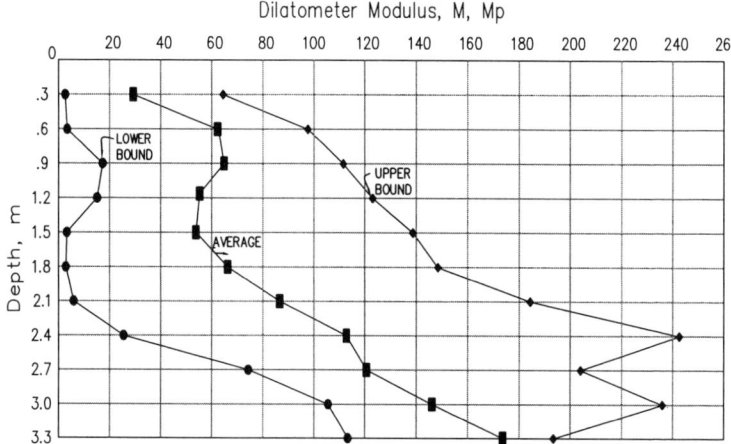

FIG. 6 Range of Dilatometer M Values With Depth

DMT soundings met vertical refusal in very firm to dense sands at depths ranging from approximately 2.5 to 5 m. Soils which are dense enough to cause DMT refusal are typically not a great settlement concern. However, the size of the lighthouse foundation required analyses of stress increases and settlement in these deeper, denser soils. To estimate elastic modulus values at depths beyond DMT refusal, LAW developed a site-specific correlation between the SPT N-values and the DMT modulus. The correlation indicated that, for analysis purposes, the M-value in Mp could be taken as $120\log(N)$.

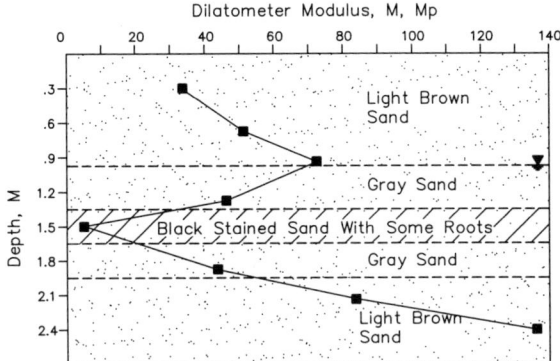

FIG. 7 Dilatometer M Values at a "Loose" Zone Area

Geotechnical Evaluation

The primary zone of concern with respect to settlements during the move was the upper loose sands. Improvement of the loose sands above the water table could be done using vibratory compaction, and this was included in the project specifications. Below the water table, there would still be a relatively thin zone of loose to firm sand about 2 to 3 m thick before the firm to dense sand was encountered. Included within this zone on some sections of the move route were the very thin, very loose sands with some organics described earlier.

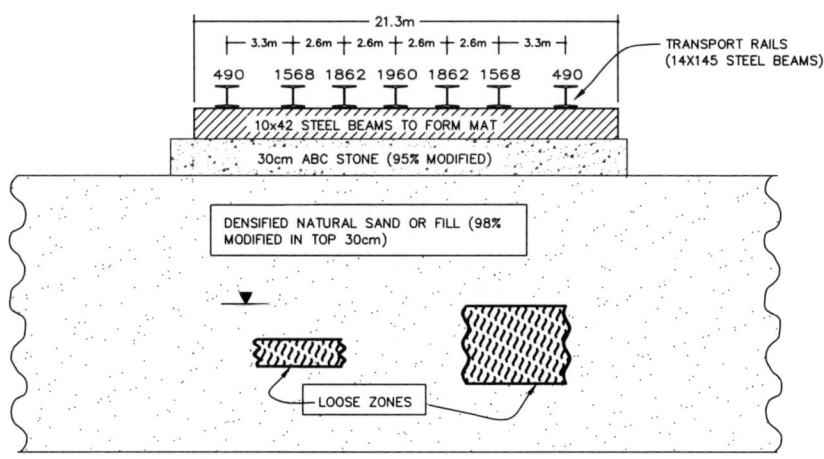

FIG 8. Design Section For Move Route

A 0.3 to 0.6 m thick layer of compacted aggregate base course (ABC) over the compacted sand subgrade was planned for placement to provide a firm, level surface for support of the steel mats upon which the seven track beams would be placed. Fig. 8 shows the final move route design and the estimated track beam loads. Fig. 9 shows the distribution of jack loading points on the two outer (edge) beams and the center beam.

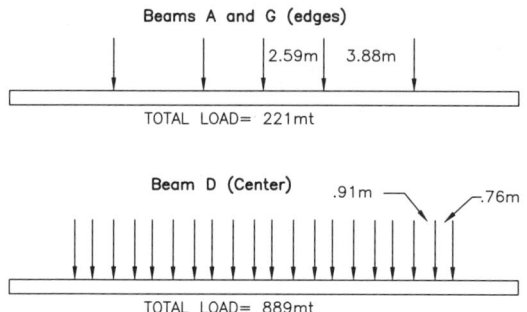

FIG. 9 Hydraulic Jack Distribution Along Track Beams

The point loads from the jacks are distributed through the track beams and further through an underlying mat formed from H-beams welded together into 1.5-m wide sections. The steel mats, the compacted ABC and the compacted soil form a stiff upper layer that reduces stresses transmitted to the deeper sands. For analysis purposes, the relative stiffness of the steel and ABC was ignored. Settlements were conservatively calculated by treating the load applied to the soil as a flexible load distributed evenly over the ground surface.

The geotechnical evaluation focused on settlement of the sand and the potential influence of the thin soft/loose layers encountered in some of the borings. Elastic layer methods using the dilatometer modulus (M-value) were used to calculate settlements. The loading was modeled as strip loads with a uniform pressure. For analysis, a 1.5 m strip 21.3 m long (the width of the steel mat) was assumed with a uniform pressure of 143 kp to 163 kp. The analyses indicated potential for settlements of about 50 to 63 mm in the center of the strip and about 25 to 38 mm at the edge. The movers indicated such settlements could be easily accommodated by the zoned jacking system.

To check the validity of using the DMT and elastic layer methods for settlement calculation, the large-scale field test described earlier was used. The test created a conical-shaped mound with heights of 1.5, 3 and 4.5 m. Elastic theory was used to calculate the expected settlement under the center of the pile using the DMT M-value profile derived from DMT soundings performed earlier at the mound test location. After reaching each incremental height, settlement of the original ground at the center of the pile was recorded. Table 1 summarizes the calculated settlements and field-measured values. The agreement between the calculations and field measurements was extremely close. The results gave confidence in the use of the M-values and elastic layer methods to calculate settlements for the lighthouse relocation.

Table 1. Comparison of Calculated and Measured Settlements for Large-scale Sand Pile Load Test

SURFACE STRESS AT CENTER OF LOADED AREA, kp	CALCULATED SURFACE SETTLEMENT, mm	MEASURED SURFACE SETTLEMENT, mm
25	3.5	4.8
50	8.6	7.9
75	15.2	16.3*

*After 12 days; 11.2 mm at initial loading

One benefit of having good confidence in the sand settlement was an ability to evaluate the thickness of compacted ABC needed. The project budget had been based on using 61 cm of ABC. Calculations indicated that reducing the thickness of the compacted ABC to 30 cm would have a negligible effect on the settlement.

The data from the additional DMT soundings performed during the construction were used to calculate the effect of the very thin, very loose zones. These additional calculations indicated the potential for an additional 19 to 25 mm of settlement in two local areas where the zone was thickest. After discussions with the movers, it was determined that the additional estimated settlement posed no significant concern for the move system.

Move Route Preparation

The move route was lightly vegetated with grass over the first 250 m, crossed an asphalt-paved parking area on one side for about 180 m, and then proceeded through scrub woods and underbrush for the last 450 m. Topographic relief was minor; elevations ranged from about 2.1 m to 2.7 m. Man-made dunes with tops at about elevation 6 m were present on the ocean side for the first 70 m. Some excavation into the land side of these dunes was required; the dunes were restored after construction. The planned move route grade began at elevation 2.4 m at the existing location and rose gently to elevation 2.9 m at the new location. At its new location, the lighthouse is about 0.6 m higher.

The move route preparation began with clearing of vegetation and removing the asphalt pavement. The clearing was restricted to a 33-m wide corridor designated by the National Park Service. A very thin surficial layer of sand with roots was stripped and stockpiled.

To provide construction access and an acceptable roadbed on which to transport the keeper's houses and other small structures, a 10-m wide roadway on the edge of the move corridor was first prepared by nominal leveling, then placement and compaction of 15 cm of ABC. The remaining 23 m of the move corridor was later prepared as the move route for the lighthouse transport.

Proofrolling of the 23-m wide move route was conducted using a 4-wheel proofroller cart typically used by the North Carolina Department of Transportation for evaluating highway subgrades. The cart, loaded with stone, weighs approximately 45 mt. The goal of the proofrolling was to aid in identifying locations of the thin, loose sand zones with organics so they could be removed or further evaluated. Although one such near-surface zone was identified and removed, the proofroller generally was not successful at identifying loose zones that were deeper than about 0.75 m. As a result, the additional DMT probes and evaluation program discussed earlier was implemented to address the loose zones and their potential impact on the move performance.

Densification of the upper loose sands above the water table was performed using a vibratory roller with a 1.22 m diameter drum, a static weight of 4053 kg and a dynamic force of 105 kN. Water was added to the exposed sandy soil to aid compaction. The compaction continued until the upper 30 cm of the sand was compacted to at least 98 percent of the modified Proctor maximum dry density (ASTM D 1557). Tests for compaction control were performed using the sand cone method (ASTM D 1556) and the nuclear gauge method (ASTM D 2922).

After the subgrade soils were compacted, the 30 cm of ABC was spread (in 2 layers) and compacted to 95 percent of the modified Proctor maximum dry density.

Because the site location is far from rock quarries, the cost of bringing in ABC was relatively high. To reduce quantities used, only sufficient ABC to cover about ¼ of the move route was brought to the site. After the lighthouse crossed over the first part of the move route, the move route components (steel mats, track beams and ABC) were picked up and moved ahead to the next section of the route. The ABC was only stripped to a depth of about 25 cm to minimize sand contamination. With reuse, the aggregate particles did have some breakage; however, the breakdown did not cause difficulty in achieving compaction or in achieving the intent of a dense, essentially non-yielding subgrade for the steel mats.

To make the transition from the original foundation at elevation 0.3 m to the move route at elevation 2.3 m, a series of steps was built by compacting aggregate base course in layers about 0.3 m thick and 2.3 m wide.   The steps were topped with the steel mats used for the rest of the move route. The track beams were supported on the steel mats by timber cribbing. The confinement provided by the ABC provided protection against localized bearing capacity deformations of the steps into the sands. A similar step arrangement was used at the transition from the move route on to the new foundation. Fig. 10 shows the step construction concept leading on to the new shallow concrete mat foundation.

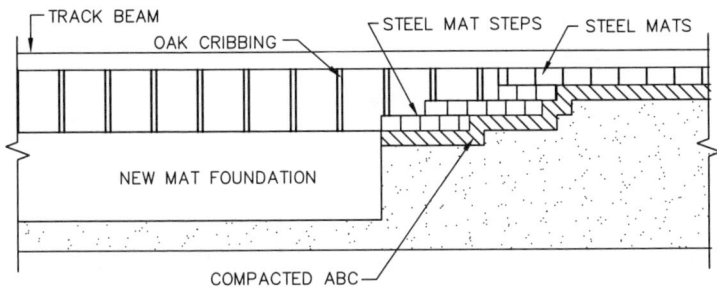

FIG 10. Transition From Move Route to New Foundation

Ground Movements During Move

High precision surveying techniques were used to monitor points on the steel mats as the lighthouse moved. During the first 24 m of the move, which brought the lighthouse off its old foundation, across the steps and on to the move route, the top step mat settled about 30 mm, less than had been predicted for the move route. This area did receive more preparation and compaction than the rest of the route, so settlements further along were expected to be closer to the predicted values.

On the third day of the move, a grid of 42 points was marked on the steel mats between the track beams, and initial readings were taken before the lighthouse reached the area. The lighthouse was stopped directly over the area overnight, and the points were surveyed again the next morning, before moving began. Figs. 11, 12 and 13 show the settlements measured. As shown in Fig. 11, a general bowl-shaped settlement pattern was seen with a maximum settlement of 28 mm in the center and less than 6 mm at the edges. Fig. 12, a section along the move route direction, shows that the maximum settlement occurred near the back of the lighthouse. Fig. 13 shows settlements across the move route at two locations. The figure shows that the maximum settlement was near the centerline of the lighthouse. The pattern of settlement matched the analysis expectations and confirmed that the support system behaved as a flexible system. The measured settlement was about half of the calculated values.

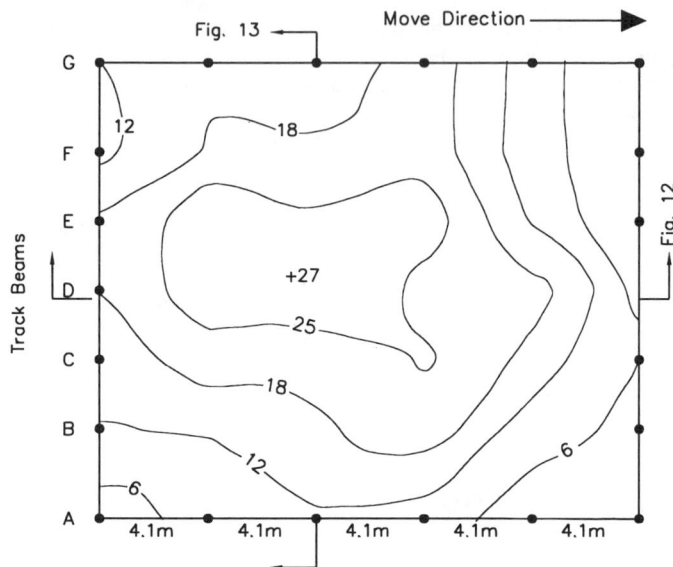

FIG. 11. Contours of Settlement at Location Near Start of Move

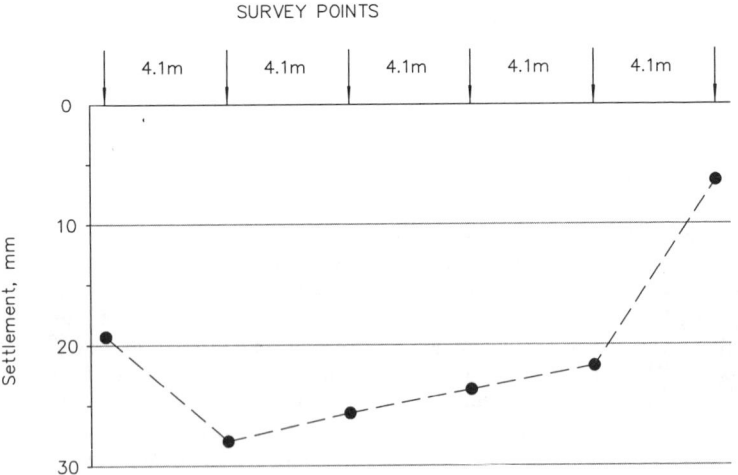

FIG 12. Settlement Under Approximate Centerline of Lighthouse

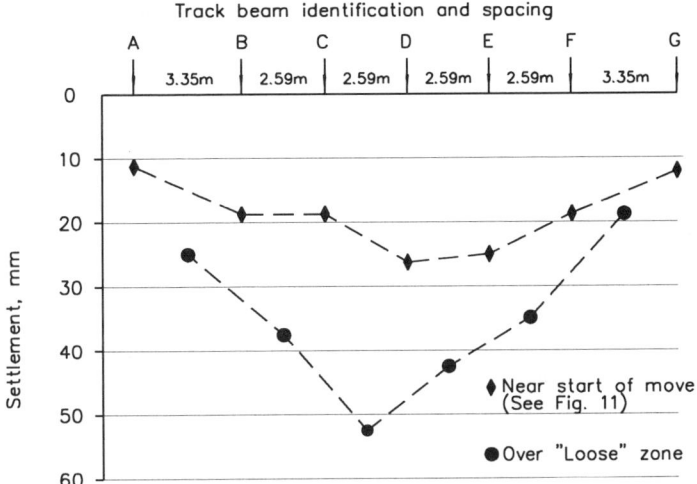

FIG. 13 Settlements Across Move Route

A location along the move route where one of the loose zones was known to be present was also monitored. Several points on the steel mats in line across the move route were surveyed before the lighthouse arrived, then after it had sat over the location overnight. As shown in Fig. 13, these readings indicated a center settlement of about 50 mm and the general bowl-shaped pattern. During the move, the settlements did not cause problems, and few jack adjustments were needed. The actual settlements were about 50 to 70 percent of the predicted values.

Conclusions

The analysis performed for this project demonstrates that settlement behavior of sands under large area loads can be conservatively modeled using confined modulus values from the dilatometer test and elastic layer analyses. By taking into account the stiffness of the steel mats and the ABC layers, it is probable that the predicted settlements would more closely match the actual settlements.

The ability of hydraulic jacking support systems to protect sensitive structures from damage during a move was clearly shown

The success of The Cape Hatteras Lightstation Relocation project also demonstrates that historic coastal structures can be safely relocated to less vulnerable sites. Stabilizing seashores is becoming recognized as a difficult, if not inefficient and ineffective, method of structure preservation. The relocation alternate is clearly feasible when seaside structures are endangered by coastal erosion.

Acknowledgements

The Cape Hatteras Relocation project was funded by the National Park Service and administered through the Denver Service Center. The Project Team consisted of:
- International Chimney Corporation, Inc., Buffalo, NY,
- Expert House Movers of Maryland, Sharptown, MD,
- DCF Engineering, Cary, NC,
- Law Engineering and Environmental Services, Inc., Kennesaw, GA and Raleigh, NC,
- Wiss, Janney, Elstner Associates, Inc., Northbrook IL and Princeton, NJ,
- Quible & Associates, Kitty Hawk, NC, and
- Seaboard Surveying and Planning, Kill Devil Hills, NC.

Special thanks are given to Dr. John Schmertmann of Schmertmann and Crapps in Gainesville FL for his service as geotechnical peer reviewer.

Pile Construction Issues at the P-700 Aircraft Carrier Wharf Project

By Mark R. Tufenkjian,[1] Arthur H. Wu,[2] and Kenneth Woehler[3]

Abstract: The paper summarizes the results of pile installation during construction of the $50 million P-700 Aircraft Carrier Wharf. The project was undertaken by the Navy to homeport a Nimitz-class aircraft carrier at the North Island Naval Air Station in San Diego, California. The existing berthing facilities were modified by creating a new land reclamation area, a new containment rock dike, and a new wharf. The wharf is supported by 450 concrete and steel pipe piles driven through the rock dike and into a dense bearing layer. A major concern during design and construction was the drivability of the piles through the rock dike and into the dense bearing layer. Results of the indicator and production pile programs are presented as they relate to construction issues. A discussion of the construction sequencing and construction difficulties encountered during pile installation are also discussed.

INTRODUCTION

The P-700 Aircraft Carrier Wharf project was undertaken by the Navy to homeport a new Nimitz-class aircraft carrier (USS John Stennis, CVN-74) at the North Island Naval Air Station. In order to accommodate these newer deep draft carriers, the North Island berthing facilities were modified to include a deepened turning basin, a new 13-acre land reclamation area contained on one side by construction of a new underwater rock dike, and construction of a new pile supported wharf. This paper focuses on the construction aspects of the pile installation for the

[1] Assistant Professor of Civil Engineering California State Univ., Los Angeles
[2] Consultant, Geotechnical Engineer Naval Facilities Engineering Service Center, Washington Navy Yard
[3] Project Quality Control Systems Manager, Nova Group, Incorporated

wharf. Other aspects of the project have been addressed elsewhere (Wu et. al., 1995; Wu and Hurley, 1997; Wu et. al., 1999; Alcorn, 1998, Schmeltz et. al., 1998). A photograph illustrating the wharf and reclamation area during construction is shown on Figure 1.

FIG. 1. P-700 Wharf and Reclamation Area during Construction

BACKGROUND

A geotechnical investigation for the project revealed that the seafloor soils in proximity to the proposed dike consisted of bay deposits underlain by beach/channel deposits, which in turn were underlain by the Bay Point Formation. In general, the bay deposits consisted of soft silty clays and loose silty sands about 3 to 7 feet in thickness with SPT blowcounts on the order of 5 blows per foot. The beach/channel deposits typically were loose to medium dense silty sands ranging in thickness from about 5 to 17 feet with SPT blowcounts from 7 to 42 blows per foot. The Bay Point Formation consisted of medium dense to very dense silty sands more than 100 feet in thickness with SPT blowcounts typically greater than 30.

Dike Design

The designers of the dike determined that the bay and beach/channel deposits were unsuitable for foundation support and had to be dredged. The over-excavated soils beneath the dike were then backfilled with quarry run prior to dike and reclamation construction. The quarry stone sizes specified for the rock dike were between 0.25 inch and 12 inches. The dike and reclamation area backfill were then constructed in multiple lifts such that a continuous slope on the outboard perimeter of the dike was maintained. The outboard slope of the dike was on the order of 1.75:1 (horizontal:vertical). The dike was designed and constructed using a multiple lift configuration to facilitate construction sequencing and to minimize the amount of quarry run needed. Armor stone was placed as shore protection on a portion of the outboard face of the dike. The average size of the armor stone ranged between 12 inches and 24 inches. Figure 2 illustrates a schematic of the multiple-lift sequencing used during construction.

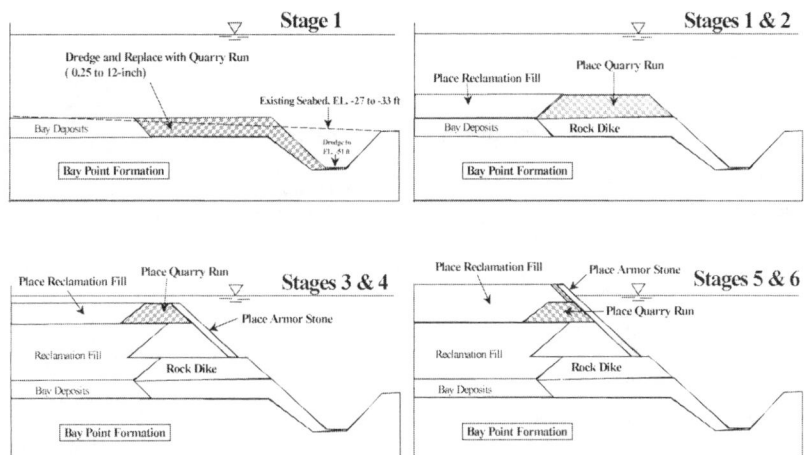

FIG. 2. Multiple-Lift Dike Construction Sequence

Wharf Design

Construction of the pile-supported wharf began immediately following completion of the dike and reclamation area. The wharf is located along the eastern edge of the reclamation area directly above the dike. The wharf is approximately 1,300 feet long and 90 feet wide. The wharf deck is supported by five rows of vertical piles designated as Rows A through E as shown on Figure 3. Battered piles were not used. A longitudinal pile cap that supports the wharf deck connects each of the pile rows.

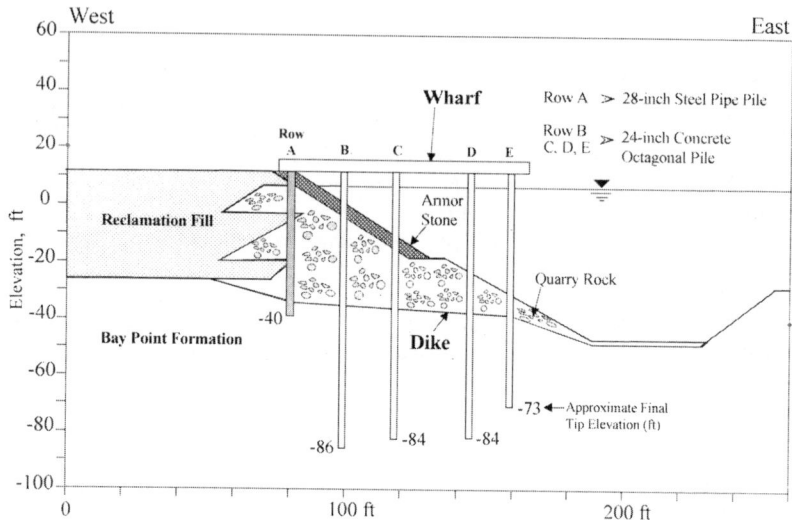

FIG. 3. Typical Wharf and Dike Cross Section

The wharf piles were designed to resist vertical and lateral loads. The piles were designed to carry the vertical loads by a combination of skin friction and end bearing by penetrating through the quarry run and into the dense Bay Point Formation. The piles in Row A were designed to resist the majority of the lateral load which could be induced by ship impact or seismic ground motion. These piles were selected and designed as 28-inch cast-in-steel-shell pipe piles (½-inch wall thickness). Piles in Rows B, C, D, and E, were selected and designed as 24-inch precast prestressed concrete octagonal piles (design prestress of 1.5 ksi).

Contract drawings indicating the minimum and estimated pile tip elevations were prepared based on structural loading of the piles and geotechnical considerations. Minimum pile tip elevations were based on pile penetration of at least 5 to 10 feet into the Bay Point Formation. In order to facilitate driving through the rock dike and Bay Point Formation, jetting (by internal jet pipes) of the concrete and steel piles was allowed to within 5 feet of the estimated final tip elevation. However, the steel pipe piles were not jetted because of the relatively short design lengths needed and the concern of disturbing the capacity of adjacent steel pipe piles due to the close center-to-center pile spacing (approximately 7 feet). The pile types and characteristics are shown in Table 1.

INDICATOR PILE PROGRAM

An indicator pile program was carried out on fourteen test piles at various locations along the length of the wharf with the primary objective of evaluating pile drivability. During the indicator pile program, the concrete piles were jetted to within 5 feet of the minimum specified tip elevation and driven unaided thereafter to evaluate driving performance. The secondary objectives of the indicator pile program were to verify design lengths and developed pile capacities with the selected pile driving equipment. Three indicator piles were driven in Row A, six in Row B, two in Row C, and three in Row D. No indicator piles were driven in Row E. All indicator piles were dynamically monitored using a Pile Driving Analyzer (PDA). Two of the indicator piles were restruck to evaluate soil set-up/relaxation.

TABLE 1. Pile Types and Characteristics P-700 Wharf Project

Pile Row	Pile Type	Number of Piles	Center-to-Center Spacing	Spacing Between Rows
A	28" Steel Pipe	181	7' – 1"	
B	24" Concrete Octagonal	59	28' - 4"	A-B: 20' - 6"
C	24" Concrete Octagonal	82	18' - 3"	B-C: 20' - 6"
D	24" Concrete Octagonal	77	18' - 3"	C-D: 30' - 0"
E	24" Concrete Octagonal	51	30' - 0"	D-E: 14' - 7"
Total		450		

The steel pipe piles were driven closed-ended (conical tip) using a Delmag D46-32 diesel hammer. The precast prestressed concrete octagonal piles were driven blunt-ended using a larger Delmag D62-22 diesel hammer. Representative pile driving diagrams reconstructed from the field logs and the PDA data are shown on Figures 4 and 5. Each diagram illustrates the pile penetration versus penetration resistance and ultimate static bearing capacity as measured by the PDA in accordance with the Case-Goble formula. Also shown on the figures are the types of materials driven through and the approximate depth of each material.

Examination of Figure 4, which is a driving diagram for one of the 28-inch steel pipe piles (no jetting), indicates that the penetration through the quarry rock was achieved with blowcounts generally less than 10 blows per foot. Once the Bay Point Formation was penetrated, there was a sharp increase in blowcounts and a final ultimate capacity of about 950 kips over the last foot of penetration.

Referring to Figure 5, which illustrates the driving diagram for one of the 24-inch concrete octagonal piles (with jetting), it can be observed that there is low to negligible penetration resistance through the quarry rock and a sharp increase in resistance once the Bay Point Formation was penetrated. Indicative of the concrete pile driving diagrams was a strong correlation between where the jetting was terminated and a sharp increase in penetration resistance. Pile driving was much more difficult and time consuming after the jetting was terminated as shown by the relatively high resistances below elevation –40 feet on Figure 5. The ultimate capacity over the last foot of penetration was on the order of 1,200 kips. Restrike tests on two of the concrete piles 1 to 2 days after initial driving indicated a soil set-up of 17% and 33%.

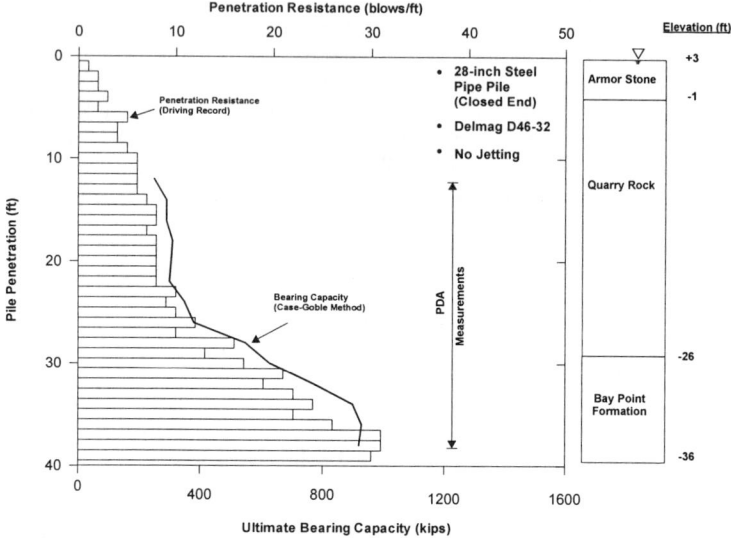

FIG. 4. Driving Diagram for 28-inch Steel Pipe Pile without Jetting.

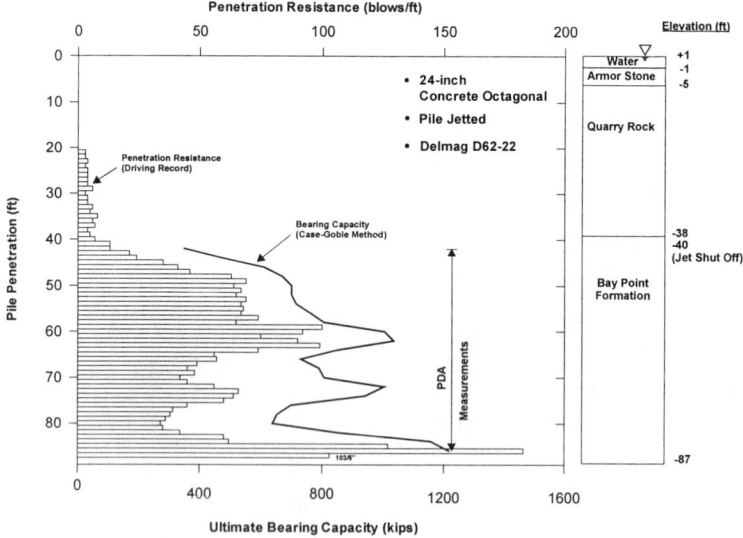

FIG. 5. Driving Diagram for 24-inch Concrete Octagonal Pile with Jetting.

The results of the indicator pile program demonstrated that the steel pipe piles could be driven through the rock dike to adequate design depths and capacities without the aid of jetting. The indicator pile results moreover demonstrated that the concrete piles could be driven through the rock dike to adequate depths and capacities; however, it was determined that jetting to within 5 feet of the estimated tip elevations would most likely be required to reach final design depths. Pile driving through the rock dike was facilitated by a combination of jetting (in the case of concrete piles), and by limiting the maximum quarry rock size. By limiting the quarry rock size to 12 inches meant that the steel and concrete pile sizes were at least twice as large as the quarry rock.

PRODUCTION PILE PROGRAM

Construction scheduling for installing the production piles was sequenced to allow the contractor to maximize the number of activities that could occur simultaneously in order to complete the project on schedule. The concrete piles were cast on-site because it was determined to be more economically feasible, it allowed greater scheduling flexibility, and it minimized the amount of truck traffic over the Coronado Bridge and through the city of Coronado to the Navy base.

A total of 450 steel and concrete piles were driven for the wharf over a 4-month period. The same pile driving equipment used to install the indicator piles was used to install the production piles. As shown in Table 2, a total of 19 additional piles were required during production pile installation, which corresponds to an increase of 4%. No additional steel pipe piles were driven and most of the additional piles were required in Rows B and C.

TABLE 2. Pile Statistics P-700 Wharf Project

Pile Row	Pile Type	No. Piles (drawings)	No. Piles (actual)	Additional Piles	% Increase
A	28" Steel Pipe	181	181	0	--
B	24" Concrete Octagonal	51	59	8	16%
C	24" Concrete Octagonal	76	82	6	8%
D	24" Concrete Octagonal	75	77	2	3%
E	24" Concrete Octagonal	48	51	3	6%
Total		431	450	19	4%

The steel pipe piles in Row A were driven from the shore using a crawler crane. The relatively close center-to-center pile spacing (approximately 7 feet) along this row facilitated production rates and on average 15 to 20 piles could be driven in a day. Pile locations were checked using surveyed offsets. The concrete piles in Rows B, C, D, and E, were driven from a crane pile-driving rig mounted on a derrick barge. Production rates for these piles were considerably slower (approximately 6 to 8 piles/day) since these piles were being driven over water and through a sloped dike.

Piles were installed in a top of slope to bottom of slope sequence (Row A to Row E) to minimize disturbance to any previously installed piles. After about 40 piles in Row A were driven (approximately 300 foot length along the wharf), pile driving for Row B was allowed to begin. Each of the pile rows B, C, D, and E, were driven approximately 100 feet ahead before falling back to drive the next row. This driving sequence allowed the piling contractor to minimize the number of times the barge was relocated and also allowed a "jump start" on the construction of the wharf deck formwork.

CONSTRUCTION DIFFICULTIES RELATED TO PILE INSTALLATION

Pile installation for the P-700 wharf project was complicated by the need to drive piles through a sloped underwater rock dike protected by armor stone. The major construction difficulty encountered during pile driving operations was maintaining pile location and plumbness during initial driving. Once the pile was set and initial driving complete, the installation proceeded relatively smoothly. Maintaining proper pile location and plumbness was important because of contract specified design tolerances and because of the contractor's customized prefabricated formwork system used to construct the wharf deck. A pile incorrectly located or out of plumb resulted in more labor-intensive formwork, which resulted in lost time and additional cost.

In order to manage the difficulties associated with initial pile driving through the sloped armor protected rock dike, the contractor tried to anticipate potential problems and counter with suitable driving methods and equipment set up. The most significant adjustment was to modify the derrick barge equipment used for pile driving. Using structural steel members, a pile driving template was constructed off the back end of the barge approximately 7 to 10 feet above the water level. The template was hydraulically driven to allow for pile location adjustments in the horizontal plane and to assist with driving the pile as plumb as possible not only during initial driving, but during driving to the final tip elevation. Pile locations were determined and checked after the pile was driven using measured offsets from the shore.

Even with the measures the contractor took to maintain design location and plumbness, it was still not uncommon for piles to have a tendency to "walk" during initial driving because of the sloped dike face and size of the armor stone protection. In some cases this required having to reset the piles several times in order to avoid other obstructing armor stone.

As the production pile operations progressed, pile driving techniques were further refined to eliminate some of the continued difficulties associated with setting and initial driving. One modification was to remove the armor stone and bench the slope at the pile location using a land-based excavator. However, because this method was done from shore, benching the slope was not feasible for piles in Rows D and E.

CONCLUSIONS

This paper summarized some of the results and construction issues related to pile installation at the P-700 Aircraft Carrier Wharf project located at the North Island Naval Air Station in San Diego, California. The wharf is supported by 450 concrete and steel pipe piles driven through a rock dike. The results of the indicator and production pile program showed that pile drivability through the rock dike and into the underlying Bay Point Formation was achieved. Pile driving was facilitated by jetting and limiting the maximum size of the quarry run to 12 inches. Piles were installed in a top of slope to bottom of slope sequence to minimize disturbance to previously installed piles. The major obstacle during pile installation was maintaining pile location and plumbness during initial driving through the protective armor stone and quarry run. Modifying the driving equipment by constructing a pile driving template, and removing armor stone and benching the dike at the pile locations, facilitated pile set-up and driving and minimized the tendency for piles to "walk" along the sloped dike.

ACKNOWLEDGMENTS

The authors would like to thank the support of the Naval Facilities Engineering Command, Southwest Division, and in particular Ms. Wendy Thornton, project engineer.

REFERENCES

1. Alcorn, A., (1998), "Environmental Impacts of Dredging to Support the Homeporting of a Nimitz Class Aircraft Carrier at Naval Air Station, North Island, Coronado California," Proceedings of Ports 98, Michael A. Kraman, editor, March, pp. 1121-1129.

2. Otus, M., Serventi, A.M., and Toda, Y., (1983), "Geotechnical Engineering, Design and Construction, Port of Oakland," Proceedings of Ports 83, Kong Wong, editor, March, pp. 129-140.

3. Schmeltz, E.J., Pierce, K.A., and Hubler, C.A., (1998), "Feasibility Study for Homeporting of Nimitz Class Aircraft Carriers," Proceedings of Ports 98, Michael A. Kraman, editor, March, pp. 1111-1120.

4. Wu, A.H., Tufenkjian, M.R., and Thornton, W., (1999), "Evaluation of Pile Monitoring and Installation P-700 Aircraft Carrier Wharf Project, Naval Air Station North Island, San Diego, California," report prepared for the U.S. Navy, Naval Facilities Engineering Command, Naval Facilities Engineering Service Center, Washington Navy Yard, Washington D.C.

5. Wu, A.H., and Hurley, B.T., (1997), "Seismic Stability and Displacement of an Embankment and Pile Supported Wharf," Proceedings of the 9th International Conference on Computer Methods and Advances in Geomechanics, Wuhan, China, Jian-Xin Yuan, editor, November, 1813-1818.

6. Wu, A.H., (1995), "Seismic Hazard Study and Geotechnical Engineering Analysis for Aircraft Carrier Wharf and Dike, Naval Air Station North Island, San Diego, California," report prepared for the U.S. Navy, Naval Facilities Engineering Command, Southwest Division, July.

Earth-Filled Wide-Base Hollow Piers
for Excavation Support

By X. Wu[1], Member ASCE, M. S. Wang[2], M. C. Wang[3], Member ASCE

ABSTRACT

Tiebacks are frequently used to support deep excavations. Installations of tieback systems require certain favorable subsurface conditions as well as skill and experience. When the subsurface conditions are not suitable for tieback installation, an alternative bracing method is needed to achieve the same degree of effectiveness without the use of anchors. This paper presents a method of using earth-filled wide-base hollow piers to brace a deep excavation for construction of Penyun Hotel Building in Shenyang, China.

The building covers 16,000 m^2 and had a 100-m high main tower with a 24-m high apron building. The depth of excavation varied from 12.3 m to 16.5 m. The foundation site contained a 4.1-m thick fill of construction debris followed by a 0.7-m thick clay, a 5.5-m thick sand, and a very thick gravelly sandy soil stratum. The ground water table was near the foundation base. The design, construction, and performance of the bracing system are presented and discussed.

It was concluded that the earth-filled wide-base hollow piers could effectively utilize a portion of unstable backfill to become a part of the retaining structure. As a result, a considerable saving in concrete material was realized. Therefore, it is an effective and economical method for bracing deep excavations when the subsurface condition are not favorable for installation of tieback systems.

---

1   Instructor, Department of Engineering Graphics, Pennsylvania State University, University Park, PA 16802, Telephone: (814)863-1537, Fax: (814)863-7229, Email: xxw101@psu.edu
2   Professor Emeritus, Department of Civil and Transportation Engineering, Northeastern University, Shenyang, China, 110006, Telephone: 011-86-(24)2391-4368
3   Professor, Department of Civil and Environmental Engineering, Pennsylvania State University, University Park, PA 16802, Telephone: (814)863-0026, Fax: (814)863-7304, Email: mcw@psu.edu

INTRODUCTION

The often used methods of deep excavation bracing include cross bracing, raker bracing, tiebacks, ring wales, and others (Dismuke, 1991). The suitability of a method to a project depends primarily on the size and depth of excavation, subsurface conditions including soil, rock and ground water elevation, loading condition surrounding the excavation area, ease of brace installation, and cost. The tieback anchor method is based on the principle of combining the retaining members such as piles, piers, and/or walls with the unstable zone of the backfill to form a composite retaining structure. Tieback anchors are perhaps the most often used bracing method. Installation of a tieback system, however, requires certain favorable subsurface conditions as well as skill and experience. When the subsurface conditions are not suitable for anchor installation, an alternative bracing method is needed to achieve the same degree of effectiveness. This paper presents an alternative excavation support method. It proposes using wide-base hollow piers to brace a deep excavation. This method was applied during construction of the Penyun Hotel in China. The construction project, and the design, construction, and performance of the hollow piers are discussed.

CONSTRUCTION PROJECT

The Penyun Hotel construction project is located in Shenyang, China. The structure covers 16,000 m^2, and has a 100 m high main tower with a 24 m high apron building. The excavation area was 88 m by 171 m, and the depth of excavation varied between 12.3 m and 16.5 m. The soil at the site was composed of a 4.1 m thick fill of construction debris followed in order by a 0.7 m thick clay, 2.5 m medium sand, 3.0 m coarse sand, and then a thick layer of gravelly sandy soil. The ground water table was near the base of foundation. The high ground water table coupled with the sandy soils below the bottom of excavation made installation of tieback anchors difficult. Because of these unfavorable subsurface conditions for a tieback bracing system, the foundation excavation was braced using earth-filled wide-base hollow cast-in-place reinforced concrete piers.

DESIDN AND CONSTRUCTION OF HOLLOW PIERS

The layout of the bracing system at the northeast corner of the excavation area is illustrated in Figure 1; a total of 103 piers are shown at this corner. Depending on the depth of excavation, the piers were built with four different lengths – 12.3 m, 12.6 m, 14.7 m, and 16.5 m. Each pier had an outside diameter of 1.8 m, an inside diameter of 1.3 m, and a wall thickness of 0.25 m. The wide base of the pier had an overall diameter of 4.8 m. The base thickness was 1.4 m at the center portion within the shaft perimeter but only 0.4 m at edge as shown in

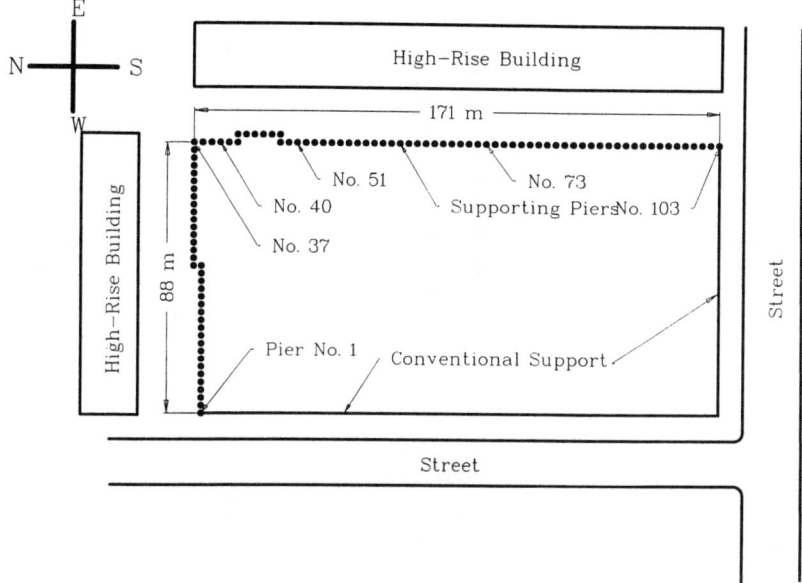

**Figure 1. The Layout of the Bracing System at
the Northeast Corner of the Excavation**

Figure 2. The hollow core was partitioned along its height into separate compartments using 0.2 m thick reinforced concrete plates. Each compartment, approximately 1.5 m high, was filled with the local sandy soil. The earth-filled wide-base pier will have a higher moment of inertia when compared with the solid pier having the same cross sectional area. Therefore, it greatly increases the capability of moment resistance. Sandy soils were used to fill the hollow cores in order to increase the weight of the piers, and therefore, increase the stability. The wide base also increases the stability because the soil above the extruded portion of the base and the pier itself acts together as a unit that increases moment and sliding resistance..

Figure 2 presents the vertical section of the finished pier dimensions. The piers were constructed manually with simple tools. The shaft was dug in steps. In each step, the shaft was advanced approximately 1 m in deep and was supported by cast-in-place concrete. The concrete forms for making the steps were slanted at approximately 15° from the vertical as shown in Figure 3, which illustrates the construction of two concrete steps. These concrete steps along the pier periphery can be used for workers to enter/exit the shaft during construction. Rapid-setting concrete was used so that the forms could be removed in 24 hours.

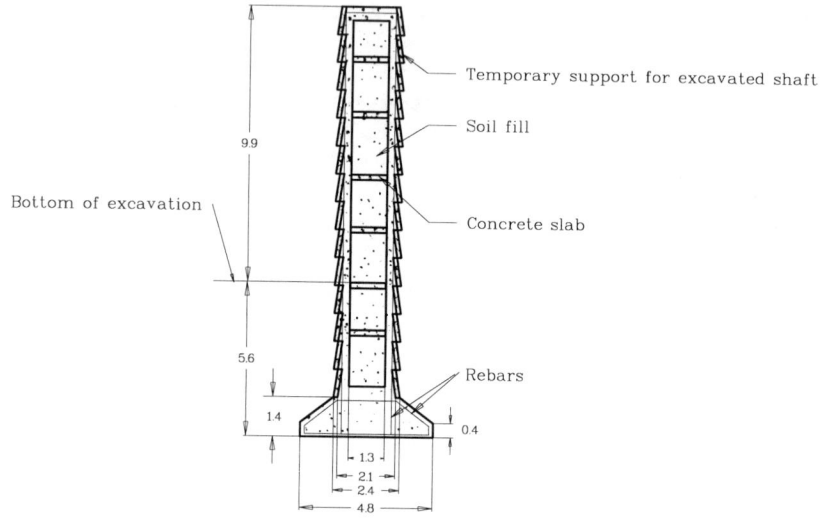

**Figure 2. Earth-Filled Wide-Base Hollow Pier Details
(Dimensions in meters)**

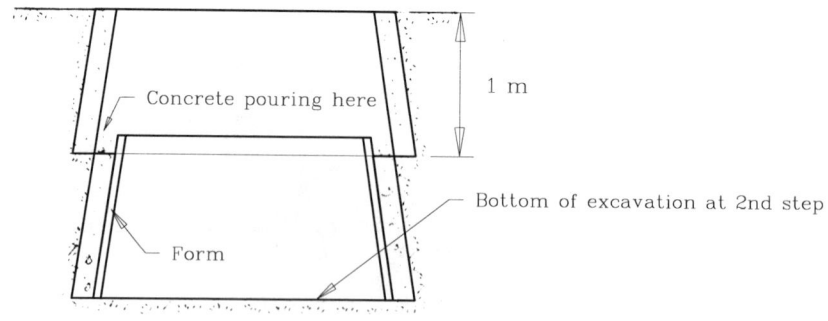

Figure 3. Illustration of Slanted Concrete Form during Construction

After the hole had reached the desired depth, the main reinforcing steel cage was lowered into the hole.   The hollow pier shaft was constructed in segments of approximately 1.5 m high.   During construction, a 1.5 m high inner concrete form was erected at bottom followed by concrete casting.   After the concrete was set, the form was removed and the inner space was filled with the local sandy soil.   The fill was then covered with a 0.2 m thick concrete slab, as shown in Figure 4.   After the slab concrete was set (hardened), the same construction procedure was repeated to build the overlying 1.5 m segments until the entire pier shaft was completed.

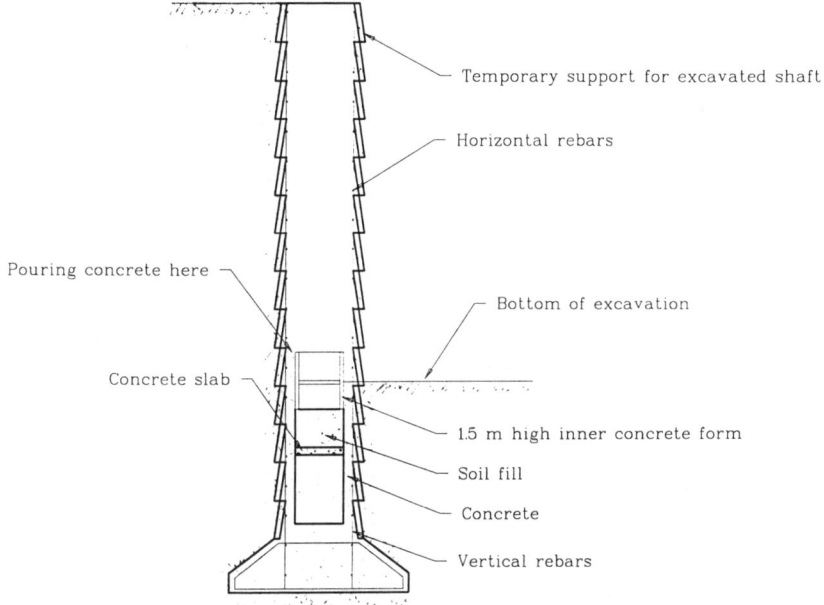

**Figure 4. Illustration of Construction of the
Earth-Filled Wide-Base Hollow Pier**

PIER PERFORMANCE

To monitor the pier performance, pressure cells and strain gages were installed on selected piers, No. 40, 51, and 73 (Fig. 1). Strain gages were bonded to the rebars at five locations; and the pressure cells were placed at four heights along the pier shaft as well as at the bottom of the base. The installation of strain gages and pressure cells, the casting of concrete, and the foundation excavation were completed on October 22, 1993, October 28, 1993, and May 21, 1994, respectively.

    The layout of instrumentation on pier No. 40 together with lateral earth pressures and base pressures monitored at three different times is illustrated in Fig. 5. The three measurement dates were October 28, 1993, November 24, 1993,

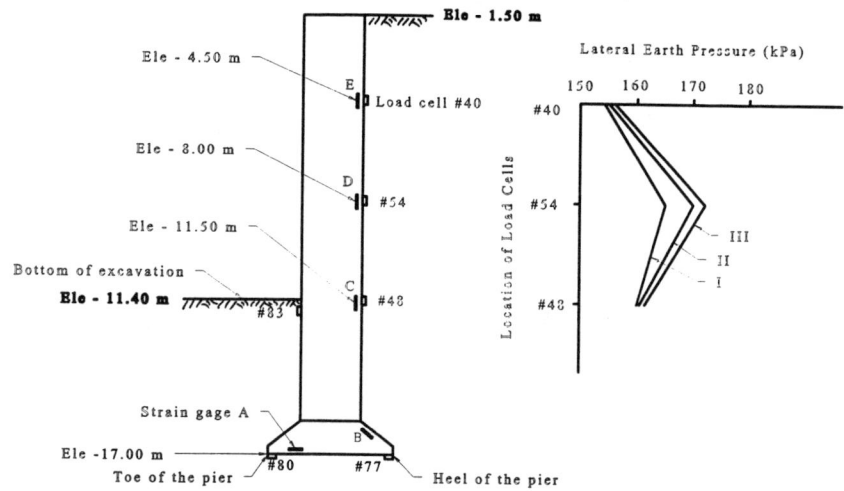

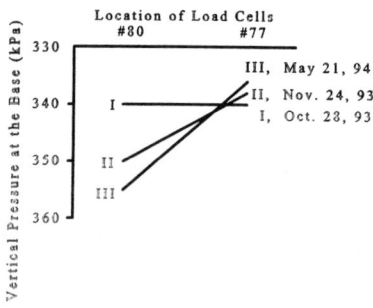

Figure 5 Layout of Instrumentation on Pier No. 40 and Test Results of Lateral Earth Pressure and Base Pressure

and May 21, 1994; the depths of foundation excavation at these three dates were 0, 5, and 13 m, respectively. As seen, the base pressure distribution was not uniform. As the depth of excavation increased, the toe pressure increased from 340 KPa (on Oct. 28, 93) to 355 Kpa (on May 21, 94) while the heel pressure decreased from 340 Kpa (on Oct. 28, 93) to 335 Kpa (on May 21, 94) as would be expected. According to the data, the base pressure was in compression throughout the entire area indicating that the resultant of the base pressure acted within the middle third of the base width. Furthermore, the lateral earth pressure increased then decreased with increasing depth of excavation. The decreased lateral pressure with depth revealed the effect of soil arching. The effect of soil arching became more pronounced as the depth of excavation increased as would be expected.

The strain gage data, which are not presented here, indicated that tension in the rebars increased with increasing depth of excavation. The percentage of strain increase was greater at top than at bottom. The tension in the rebars at bottom of the pier shaft indicated the cantilever action of the bracing system. Thus, the earth-filled hollow piers behave to some degrees as cantilever retaining structures.

CONCLUSION

For this excavation project, a bracing system using solid piers was also analyzed and evaluated. According to the analysis, the required solid pier dimensions were 1.23 m in diameter with a 10.35 m penetration depth. Because of the smaller diameter, compared with that of hollow piers, more solid piers than hollow ones would be required in order to maintain the same clear pier spacing for soil retention. As a result, considerably more concrete material would be needed for construction of solid piers than for hollow piers. Since the water table was near the bottom of excavation, and the lower portion of the solid pier would have been under the water table, another disadvantage of the solid pier bracing system was the need to construction under water. To avoid working in water, the excavation site would have had to be dewatered. Dewatering would have added considerable cost to the overall budget.

Major advantages of earth-filled hollow piers include: 1) the construction takes place above the ground water table, and 2) it effectively utilizes the unstable portion of the backfill above the extruded portion of enlarged base. This unstable portion of the backfill with pier, including the soil inside, act together to form a composite structure that behaves like a gravity retaining wall. In other words, it effectively utilizes the unstable portion of the backfill, resulting in a considerable saving of concrete material. However, hollow piers are not without weaknesses. The major weakness perhaps is in the construction of the hollow core and the enlarged base. The hollow core construction requires erections of inner forms that can be labor intensive and costly. For the construction of the enlarged base, it requires extra precautionary measures to avoid soil caving and collapsing.

The shape of earth-filled hollow piers mimics that of floating caissons. However, there are fundamental differences between the two. The hollow piers are cast-in-place, whereas, the floating caissons are prefabricated elsewhere, and then transported and sunk at the site. The size of hollow piers is considerably smaller than the normal size of floating caissons. The core space in the hollow piers is partitioned horizontally into segments, while that of floating caissons is divided, if any, into multiple cells using vertical partitions (Jumikis,1971). The base to shaft diameter ratio is greater for hollow piers than for floating caissons. With this large base to shaft configuration, hollow piers are able to mobilize a large volume of unstable backfill to form an effective retaining structure.

REFERENCES

Dismuke, T. D. (1991). "Retaining Structures and Excavations". Chapter 12 of Foundation Engineering Handbook, H. Y. Fang, Ed. Van Nostrand Reinhold, New York.

Jumikis, A.R. (1971). Foundation Engineering. Intext Educational Publishers, Scranton.

CONSTRUCTION CLAIMS DUE TO DIFFICULT MUCK EXCAVATION AT THE TWIN LAKES DAM ENLARGEMENT PROJECT

Theodore B. Feldsher, P.E.[1]
Daniel L. Johnson, P.E.[2]

Abstract

The Twin Lakes Dam Enlargement project, located in Northern Wyoming, was completed in 1998. Foundation excavation for the embankment dam required removal of deep peat bog deposits. The excavation depths and bid quantity were estimated based primarily on hand probe data, with an additional amount added to allow for uncertainty. Due to strong competition, the successful bid for construction was substantially lower than the Engineer's Estimate. The low bidder planned to complete the work in a single construction season, about half of the duration estimated by the Engineer. During construction, much deeper than expected peat bog deposits were encountered. This impacted the schedule, and the Contractor eventually took three full seasons to complete the project. Ultimately, the Contractor negotiated additional compensation for increased costs associated with the muck excavation, including delay impacts and inefficiencies. However, even with large increases from the low bid price, the final construction cost was within 10 percent of the original budget estimate.

Introduction and Background

The Twin Lakes Dam and Reservoir Enlargement Project was constructed by the Sheridan Area Water Supply Joint Powers Board, between May 1996 and November 1998. The reservoir enlargement was built as part of a larger project to increase the water supply for the City and County of Sheridan, Wyoming. The project site is located at an Elevation of 8600 feet in the Big Horn Mountains, about 23 miles southwest and about 4500 feet higher in elevation than the City of Sheridan (pop. 15,000). The site was first developed for water supply in 1937, when the City constructed a 54-foot high dam to retain Twin Lakes Reservoir No. 1, with a storage capacity of about 1300 acre-feet. In 1969, the City constructed a 22-foot high dam to retain Twin Lakes Reservoir No. 2, which stored an additional 200 acre-feet. Before the reservoirs were constructed, small natural lakes, peat bogs, and wet meadows were present at each site.

---

[1] Senior Engineer, URS Greiner Woodward Clyde, Oakland, CA

[2] Principal, URS Greiner Woodward Clyde, Denver, CO

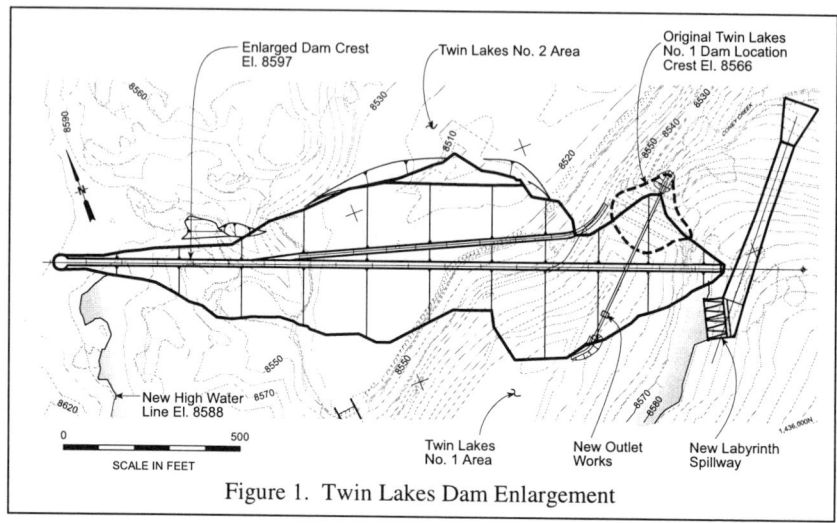

Figure 1. Twin Lakes Dam Enlargement

The enlargement project consists of a new embankment dam, spillway, and outlet works, as shown in Figure 1. The existing two-tiered reservoir system was replaced with a single larger reservoir, with a new storage capacity of about 3400 acre-feet. The old No. 2 reservoir was breached and the area restored. The new dam has a crest length of about 1,800 feet, a maximum structural height of about 125 feet, and was constructed as a zoned earthfill embankment, shown in Figure 2. A new labyrinth weir spillway cuts through bedrock at the right abutment. The outlet works consists of an intake tower with multiple level inlet gates, an outlet conduit passing beneath the dam, and an energy dissipation structure at the downstream toe. The Owner retained Woodward-Clyde in 1990 to perform geotechnical site investigations, to prepare designs for the dam and appurtenant facilities, to prepare construction contract documents, and to provide resident engineering and design services during construction.

The original project configuration first proposed in the late 1980's included a reservoir capacity enlargement up to 4600 acre-feet. That project would have combined and raised both existing reservoirs with a single new dam built along a downstream alignment. Site investigations were undertaken for such a dam, and final designs were prepared. However, before the project advanced to construction, the U.S. Army Corps of Engineers and the U.S. Environmental Protection Agency denied key permits, citing unacceptable inundation of sensitive wetlands in the proposed reservoir area. As a result, the Owner reconfigured the project to reduce the impacts, reapplied for the permits, and the revised design was eventually accepted. The new design reduced the amount of impacted wetlands to about 6 acres, compared with about 22 acres in the original design. To mitigate these impacts, the redesign included breaching the outlet of Twin Lakes No. 2, to revert it back to a natural lake. This allowed restoration of about 8 acres of wetlands formerly inundated by the reservoir.

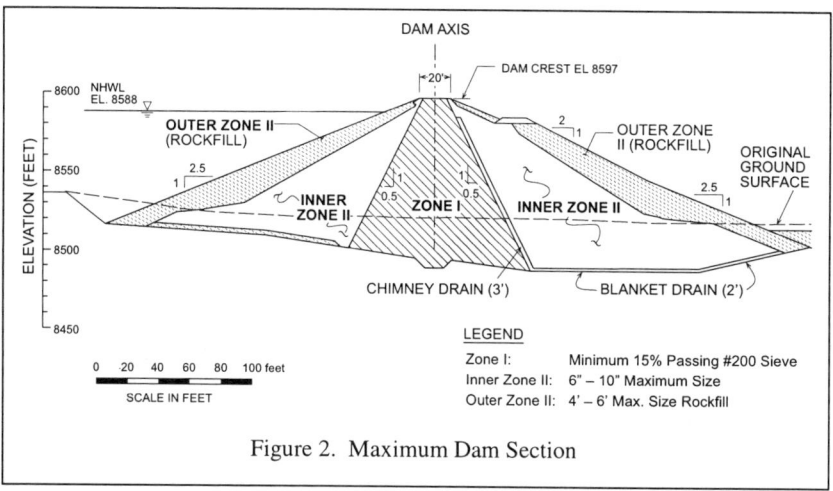

Figure 2. Maximum Dam Section

These restored wetlands more than made up for those impacted by the new dam and reservoir, and as a result the enlargement project created a 2-acre net gain in the total wetlands area at the site.

The redesign spared the most sensitive areas, but decreased the storage capacity of the enlargement project by about 1200 acre-feet and further reduced the already low storage efficiency. In order to maximize the remaining reservoir storage capacity, the redesign pushed portions of the new embankment footprint into a peat bog area at the upstream end of the No. 2 reservoir. Relatively little was known about foundation conditions in this area, since access for subsurface investigations was difficult. However, since significant uncertainty remained at the time about whether the revised project would be accepted by the permitting agencies, and since the cost efficiency of the enlargement had already been reduced, both the Owner and Engineer were somewhat reluctant to undertake an expensive program of additional difficult-access investigations. The Owner had already funded investigations and design for the original project, and so the Engineer utilized the prior data about the peat bog depths for the new design, along with some relatively economical hand probe investigations. As described below, this led to an incorrect estimate of the depth and quantity of muck excavation required for the new dam. This in turn gave the Contractor a basis for several large claims and change orders, which led to a final contract price much higher than the original low bid.

Site Conditions and Investigations

The Big Horn Mountains were extensively glaciated during the late Pleistocene era, and much of the project area is covered with the associated moraine and till deposits. These include sands, gravels, and cobbles, often in a silty matrix, with many boulders greater than 5 feet in diameter. Precambrian-age granitic bedrock outcrops are present in the right abutment and spillway chute areas. Deep deposits of the glacial till underlie most of the rest of the enlarged dam area. The topographic low areas at the site, including the bottoms of both existing reservoirs, are filled with thick deposits of peat, organic silts, other very soft glacial lake sediments, and boulders. For simplicity, these soft peat and silt deposits are referred to herein as "muck."

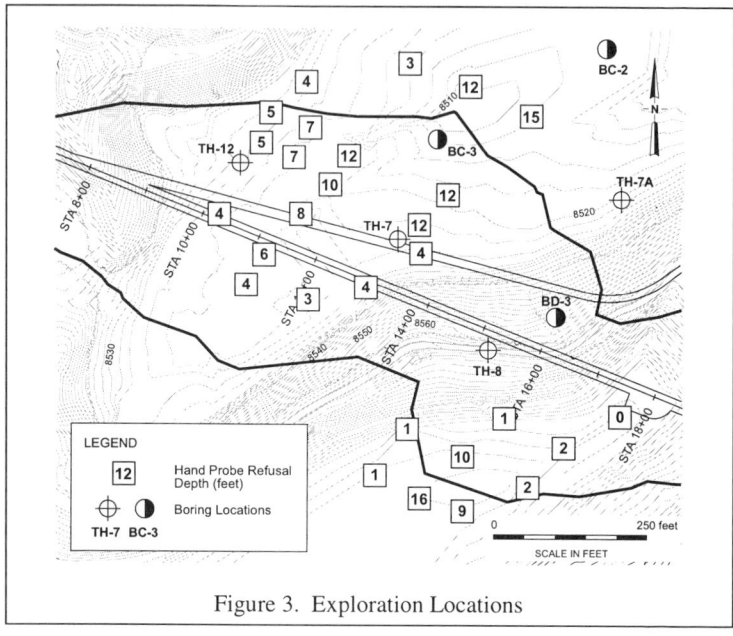

Figure 3. Exploration Locations

Various geotechnical investigations were carried out at the site between 1987 and 1995, as the project more than once developed from conceptual to final design and the project configuration evolved as noted above. The Engineer carried out subsurface geotechnical investigations at the site in 1991, 1994, and 1995, and other consultants carried out earlier investigations. The most recent work included exploratory borings, backhoe test pits, and seismic refraction surveys. Since the site is located in a popular National Forest recreational area, the amount of disturbance allowed for exploration was limited. Truck-mounted and all-terrain drill rigs were used in areas where access was feasible, and a small, portable drill rig moved by helicopter was used for the most difficult to reach locations. This equipment was not feasible for exploration of the peat bog areas within the new dam footprint. The peat areas were normally inundated, and even when exposed were generally too soft to support foot traffic, much less drilling equipment.

Consequently, the Engineer decided to explore the peat bog areas with hand probes, which are a somewhat crude (but not uncommon) method for roughly assessing the depth of very soft deposits. The probing was done by manually driving thin steel rods down into the peat until refusal was reached, and recording the depth of refusal. Access for driving the hand probes was obtained by boat in some areas, and from atop a winter ice sheet in other areas. A total of about 28 probes were driven, 19 of which were located in the Twin Lakes No. 2 area, in the vicinity of the maximum dam section. As shown in Figure 3, the probe refusal depths varied gradually over the area, and ranged up to 16 feet. Nothing in the refusal depth data suggested that soft materials were present to greater depths, nor did the data suggest the presence of boulders or other erratics within the soft deposits.

Even though the probe refusal depth data were consistent, the Engineer recognized that uncertainties remained, since no samples were retrieved. Although the refusal material was expected to be the firm glacial till, the only way to verify this would have been to mobilize a barge to the site and conduct drilling operations over water, or to mobilize a drill in wintertime and use the ice for access. These both would have been difficult and expensive undertakings, and were decided against. This decision was based in part on the existence of prior borings from a 1989 investigation by an earlier project consultant. Those borings were located in the Twin Lakes No. 2

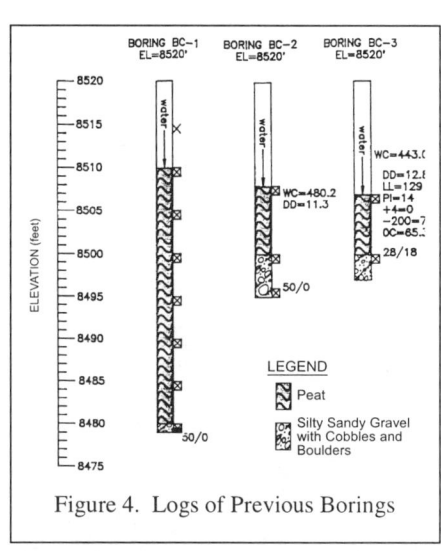

Figure 4. Logs of Previous Borings

area, downstream of the new embankment toe. They had been drilled during studies for the earlier project configuration (which had a different dam location). Although the logs are limited in detail as shown in Figure 4, the two closest borings (BC-2 and BC-3) clearly showed muck depths of 14 to 16 feet over much denser materials. These results were consistent with nearby hand probe refusal depth data, and so the combined data was judged reasonable for estimating the depth of the muck deposits. For bidding purposes, the muck excavation quantity was estimated based on the probe refusal depths plus 5 feet, to allow for an additional margin of uncertainty. With this 5-foot allowance, which substantially increased the estimate, the final muck excavation bid quantity was set at 79,000 cubic yards.

Contract Bidding Phase

In order to begin construction in the spring of 1996, the project was advertised for bidding in the fall of 1995, and a preconstruction site visit was conducted before the site was snowed in for the winter. Many contractors showed interest in the project, and the Owner received a total of five bids on the April 1996 bid date. The relatively high level of interest in the project resulted in strong competition between bidders and correspondingly low bid prices. This situation is ordinarily desirable from an Owner's standpoint, but in this case the very low price also carried significant negative repercussions, in terms of increased numbers of disputes and claims during construction.

The five bids submitted ranged from a low of $6.9 million to a high of about $11 million, as summarized in Table 1. The three lowest bidders proposed construction schedules are also summarized, showing that their prices are roughly proportional to their schedules. The Engineer's cost and schedule estimates are also presented for comparison.

Table 1. Summary of Bids and Schedule Durations

Bidder	Total Bid Price	Construction Schedule Duration
A	$6,855,658	Finish 4 months early
B	$7,317,000	Finish 1.5 months early
C	$9,379,022	Two full construction seasons
D	$9,924,040	--
E	$10,963,670	--
Engineer's Estimate	$10,608,785	Two full seasons of construction (only if no problems encountered)
	$11,139,224	Three full seasons of construction (recommended for budgeting)

The Engineer prepared his construction schedule estimate for the project based on the limited time available for work at the site each year, given the high elevation and winter snowpack. Vehicle access to the site is typically feasible only from late May through early November, and snow removal can be required for access both at the beginning and end of the season. The Engineer estimated that at least two full construction seasons would be needed for the work, assuming weather conditions were favorable and no major unforeseen conditions or delays were encountered. Since such conditions and delays are common, the Engineer recommended that the Owner budget to allow for a third construction season, to cover likely costs of additional mobilization/demobilization, equipment rental, and additional field overhead.

Construction Schedule Issues

Prior to award of the contract, the Engineer and the Owner weighed the viability of the low bidder's very aggressive single-season construction schedule submittal. Given the realities of construction at the difficult and remote site, the Engineer concluded that the probability of Contractor actually completing the work in one season was low. Therefore, the initial schedule submittal was rejected. The Contractor subsequently made minor changes and resubmitted, this time showing the majority of construction still completed within one season, but with about 1-1/2 months of work in the second season. As before, the schedule still showed a large number of critical path items, potential for multiple critical paths, and what appeared to be highly optimistic time duration estimates for many items.

Although not explicitly stated, the Contractor's schedule showed about 400,000 cubic yards of embankment fill placement (60 percent of the total dam volume) between mid-September and mid-November of the first season. The corresponding average placement rate of about 8000 cubic yards per day was actually achieved for a portion of the 1997 season, using double shift placement operations, 6 days per week. Actual 1997 daily placement progress is summarized in Figure 5. Freezing conditions and heavy snowfalls are common at the site after about early- to mid- October, and fill placement rates can be significantly impacted. However, the Contractor's schedule did not appear to take this into account. During the first season of construction, the Contractor's actual mid-September to mid-November fill placement was substantially impacted by inclement weather, and totaled only about 150,000 cubic yards.

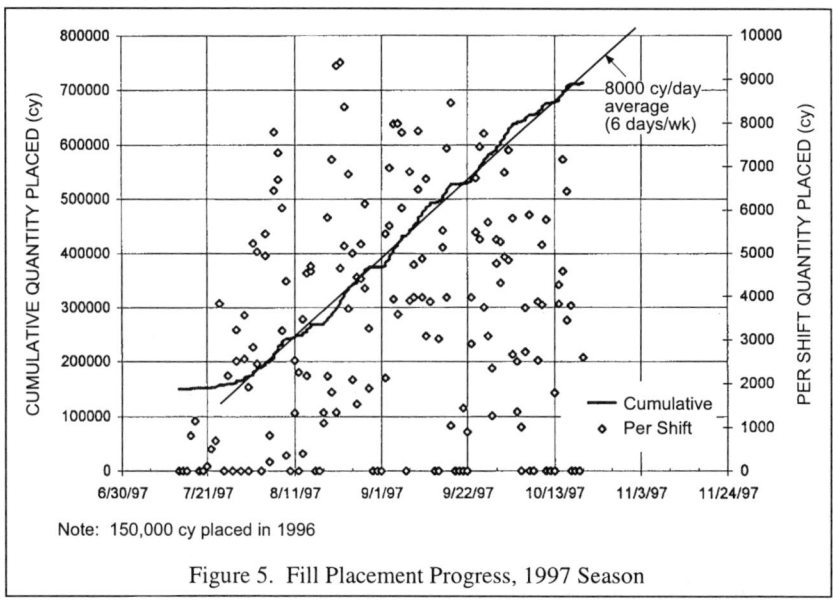

Figure 5. Fill Placement Progress, 1997 Season

Prior to award of the contract, the low bidder was understandably reluctant to make significant modifications to his schedule, because his low bid price was dependent on his projected production rates and his forecast for early completion. The Engineer remained skeptical, but after reviewing the Contractor's qualifications and experience and finding them satisfactory, recommended award to the low bidder. The Contractor's aggressive schedule submittal was neither formally accepted nor rejected.

Much later, the Contractor submitted delay claims based on his initial schedule submittals. He argued that since his low bid had been accepted, and since his schedule had not been rejected, the Engineer and Owner had implicitly accepted his aggressive plan for the work. Unfortunately, the bid documents contained no provisions to disallow such unrealistic early completion schedules. Without such protection, the door was left open for bidders to make very aggressive production assumptions and to bid low based on the resulting short schedule. By bidding on an early completion schedule, the Contractor effectively takes ownership of the float time between his early completion date and the contract end date. Such bids can then provide a setup for subsequent "delayed early completion" claims once the aggressive production assumptions fall apart, even if the work still gets done within the allowed contract time. One possible defense against this problem is to require all bidders to base their price on using the full amount of contract time allowed. In this case, bid prices would probably have been higher (although more comparable), but the amount of subsequent disputes and change orders would probably have been less.

Muck Excavation

Based on the available geotechnical data, the Engineer expected the muck excavation in the new dam foundation to be relatively difficult. The material was clearly soft and saturated, and could result in difficult access for excavation and hauling equipment. In combination with the wet and loose nature of the spoils, the excavation and hauling difficulties meant that a relatively low excavation efficiency was expected. Because of this, the contract included a separate unit price bid item for the muck excavation. The Engineer's estimate was $4.00 per cubic yard, based on a duration estimate of about 45 shifts. For comparison, the Contractor's bid price for the muck excavation was $1.25 per cubic yard, and his initial schedule showed only 14 days for completion of the work. In retrospect, the Owner and the Engineer should perhaps have more seriously questioned this low bid price and short duration estimate when the bids were initially being evaluated.

When construction began in the spring of 1996, the Contractor began the foundation excavation work as soon as possible. The work was on the critical path for embankment fill placement, and so the Contractor came prepared with an ample fleet of excavators and hauling equipment. The first days of muck excavation went as expected, but within a week or so it quickly became evident that the work was going slower than planned. Progress was hampered by the very soft and wet nature of the materials, and the excavation depths were turning out to be significantly greater than predicted. In addition, the glacial till encountered beneath the muck proved to be saturated and susceptible to deep rutting and disturbance under heavy haul truck traffic. The Contractor also began to encounter very large boulders buried in the middle of the muck area. These boulders almost appeared to be "floating" in the soft sediments, and were in many cases too large to move without first breaking or blasting them into smaller pieces. The muck deposits encountered generally consisted of about 5 to 10 feet of fibrous peat at the surface, overlying very soft organic silts, and grading at depth to less organic very soft glacial silt and clay deposits. Due to its light weight, the peat could often stand on near-vertical cut faces, but areas underlain by the softest silt deposits typically experienced progressive slumping failures shortly after the soft deposits were exposed in the excavation.

In most areas, the Contractor used timber mats to support his excavating equipment on the muck. Since the glacial till beneath the muck was wet and disturbance-prone, access ramps and roadways for hauling equipment were necessary.

To avoid constant rebuilding of haul roads, he eventually began to use multiple excavators to shuttle the material. In the most difficult and hard to reach areas, the muck was passed between three separate excavators before it was loaded into the haul trucks. Eventually, after 70 to 75 shifts of excavation (including double shift work) between June 26 and September 12, 1996, the muck excavation was completed. The final survey results (summarized in Figure 6) showed excavation depths ranging from 10 to 30 feet <u>deeper</u> than had been estimated based on the hand probes and previous borings. The total excavation depths ranged up to about 45 feet. The final total muck quantity was estimated to be about 167,000 cubic yards, compared to the original estimate of 79,000 cubic yards.

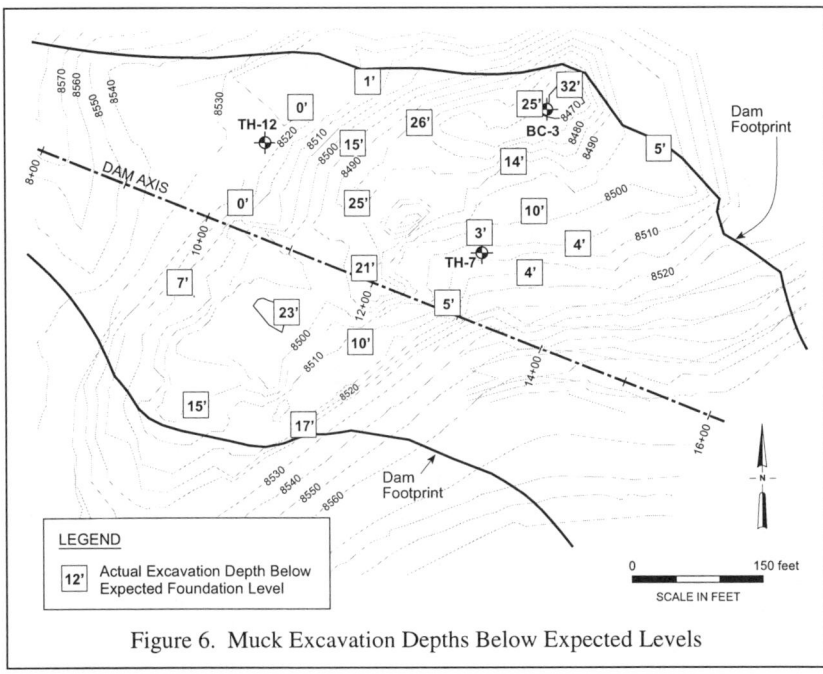

Figure 6. Muck Excavation Depths Below Expected Levels

It has yet to be determined why the data from both the hand probes and the previous borings turned out to be so misleading. It is possible that some thin dense layer could have caused the probe refusals, but such a layer was never identified in the excavation. Some other effect such as skin friction on the probe rods could have possibly have limited the maximum useful probe depth. The two existing borings (BC-2 and BC-3) could have been logged inaccurately, or perhaps may have been located incorrectly on the original map. The fact that the indicated ground surface elevations are inconsistent between the boring logs and the original map provides some support for this theory. Interestingly, the log for boring BC-1, which was located about 600 feet downstream, shows a muck depth of about 30 feet, similar to much what was actually encountered in the deepest excavation area. Unfortunately, during design this boring was probably considered too far downstream from the excavation area to be relevant.

Differing Site Conditions Claim

Not long after it first became evident that the muck excavation was going deeper than expected, the Contractor submitted an initial claim notice for Differing Site Conditions. His letter cited "delays, impacts, and inefficiencies" expected to result from the deeper and more difficult muck excavation. Initially, the Engineer acknowledged the increased depth, but was reluctant to agree with a Differing Site Condition claim since it was not yet apparent that the quantity overrun would be substantial. Later, when the magnitude of the overrun became more apparent, the Engineer acknowledged that an equitable price adjustment was justified, due to the increased difficulty of the work. Ultimately, based on various cost and production rate impact estimates, a negotiated settlement was reached. A change order was issued revising the unit price up to $3.59 for the first 79,000 cubic yards and to $6.56 for every yard over that. Both parties also agreed that some delay was attributable to the increased difficulty and quantity, but that it was too early to assess and resolve the magnitude and impacts. Therefore, the change order specifically left matters associated with Contract Time and potential delays for later resolution. In retrospect, it probably would have been better to push for resolution at the time, because it took two more years to finally settle the matter.

Deep Fill Claim

By the end of the first season of construction in early November of 1996, all of the muck excavation was complete and a substantial amount of embankment fill had been placed. The fill in the Twin Lakes No. 2 muck excavation area was completed back up to just above the original ground level. In the Twin Lakes No. 1 area, fill was placed high enough to temporarily retain the reservoir the following spring. The fill placement operations were hampered that first season by September and October snowfalls, by slow materials processing, and to a lesser degree by the fact that the fill placement areas were not yet very large. After demobilization for the season, the Contractor submitted a claim for additional compensation for the so-called "deep" fill, which referred to that extra fill required only because the deeper than expected muck excavation had to be refilled. He claimed that this extra fill was more difficult and costly than the "average" fill covered by his bid price, since it had to be placed in a relatively more confined working area with consequently lower efficiency placement operations. This and several other items were included in a claim for about $700,000. Eventually, the Contractor submitted full cost information for his fall 1996 fill placement operations, in effort to substantiate his claims. The Engineer reviewed this information and acknowledged that there appeared to be some degree of merit, but recommended a significantly lower settlement amount than claimed. The Contractor promptly rejected the recommendation as insufficient. After further discussions, the Owner and Contractor eventually reached a negotiated settlement amount of $464,000, with no time extension. This amount included about $190,000 for justified equipment standby costs associated with the extended duration of the muck excavation work.

The balance was for efficiency impacts on placement of about 60,000 cubic yards of fill material.

Delay and Disruption Claims

Following resolution of the deep fill claim, the Contractor submitted a $2.24 million claim for remaining delay and disruption impacts associated with the muck excavation. The claim alleged that the muck excavation had delayed the project by 5-1/2 months, the equivalent of an entire construction season. The Contractor maintained that the dam construction would have been completed in one season, if only the muck excavation had been as originally expected. However, based on the daily construction records, the Engineer concluded that the actual delay impact was at most 2-1/2 months, which translated to a cost impact of about $500,000. The Contractor refused to accept a change order for this amount, and the issue remained in dispute for the duration of the 1997 construction season.

Meanwhile, the Contractor also submitted a $1.83 million claim for acceleration, consisting mainly of additional equipment and supervision costs, costs for double shifting the fill placement work, and related inefficiency costs. The Engineer responded that no directive had been given to accelerate, and that double shifting had been a part of the Contractor's schedule all along, so should not constitute acceleration. The Contractor argued that since he had been impacted by the muck excavation delay, but had not been granted a schedule extension, he had been forced to into "constructive" acceleration. As noted, the Engineer acknowledged some compensable critical path delays, but reasoned that the Contractor's original schedule (with 4 months of float to the required completion date) should accommodate a 2-1/2 month delay should without requiring a time extension. The Contractor strongly disagreed with this assessment, and claimed he was at least due a 2-1/2 month time extension. The Contractor asserted that he had not planned to double shift any second season fill placement, and again maintained that he was being forced to accelerate the work at great expense.

All of this discussion was further complicated by the fact that the weather at the site had been relatively dry during the spring and summer of 1996, but the same period in 1997 was rainy. The Engineer argued that the Contractor should have been planning all along for fill placement in 1997, and should have been prepared for variable mountain weather in any case. The Contractor argued that, but for the muck excavation, he would have been able to complete more of the work in good weather in 1996, and so he was unfairly impacted by having to complete a large portion of the work in 1997's wetter weather.

By the end of the second construction season in early November of 1997, the embankment fill placement was essentially complete. A substantial amount of work remained for the third season, including the labyrinth spillway structure, the outlet works control building, and site restoration. Early in 1998, the Contractor submitted a total cost claim covering all of the work he had completed to date on the project. Without presenting a detailed cause and effect analysis for the claimed extra costs, he claimed total (although unaudited) costs of about $14.2 million for the first two construction seasons, compared to Contract payments of only about $7.8 million. He requested full compensation for the difference, and cited as justification a list of alleged changes, delays, and other consequential impacts stemming from the muck excavation. He also referenced various other allegations related to extreme weather conditions, insufficient borrow materials availability, and excessive environmental mitigation costs.

In review of the claim, the Engineer noted that the Contractor had made no credible effort to show logical cause and effect links between the muck excavation work and all of the claimed impact costs. Lacking such detailed substantiation, the Engineer concluded that the claim was not a sufficient basis for recommending additional compensation. The Contractor's position on the matter was that the entire character of the work had changed, and so he should be relieved from any responsibility for risks inherent in his aggressively low bid and optimistic schedule. The Engineer maintained that the work had not in fact changed to such an extreme degree, and so the Contractor should remain responsible. Nevertheless, the Engineer still recommended that the Owner make an equitable adjustment to compensate for justifiable and substantiated impact costs, where the Contractor's own over-optimism and inefficiency were clearly not the main cause of the impact.

Through subsequent negotiations, the Owner and the Contractor eventually reached a full and final settlement of the delay and disruption claim and all other outstanding disputed issues. As a part of the settlement, they agreed on final quantities for all unit price bid items, and agreed to change the remaining 10 percent of the Contract to a lump sum basis. A change order was prepared to increase the contract price by $2.9 million, and a time extension of 125 days was also granted, which provided for the third construction season in 1998.

Table 2. Summary Of Major Change Orders (>$50k)

Change Order No.	Description	Amount
1	Muck Excavation Unit Price Increase (Direct Costs, with final quantity)	$762,000
11	Outlet Works Foundation Changes	$60,000
18	Deep Fill Premium Unit Price plus equipment standby costs – Muck Excavation and Deep Fill	$464,000
24	Spillway Structure Changes and other items	$190,000
25	Resolve Delay, Disruption, Acceleration, and all other Claims	$2,896,000
	Total of All 29 Change Orders:	$5,007,000
	Total Attributable to Muck Excavation Increase:	$4,123,000

The project was completed by the end of 1998, at a final construction cost of $11.9 million -- more than $5 million higher than the original bid price. As summarized in Table 2, about 80 percent of this increase is directly attributable to the muck excavation, deep fill, and delay and disruption change orders. Because the original bid price was so low, final price was still only about 7 percent more than the original $11.1 million estimate for three seasons of construction. The fact that the project took three seasons strongly suggests that the Contractor's original one-season schedule was in fact unrealistic. Even without the muck excavation overrun, at least two full construction seasons would likely have been needed for the work.

Conclusions and Lessons Learned

Construction of the Twin Lakes Dam Enlargement Project included a variety of surprises, disputes, and other difficulties. These challenges ultimately made the project more expensive and time-consuming than hoped, but it was built well and should serve the Owner for many years into the future. Some of the more important and perhaps generally applicable lessons learned include the following:

- Even though the Engineer may follow the usual professional standard of care in reaching a conclusion about subsurface conditions, he or she is likely to be held accountable if something unexpected is found instead.

- The Engineer's only defense in some such situations may be the fact that the Owner did not hire them to provide a "perfect" design. Error-free perfection would be far more expensive than most Owners are willing to pay for, and is generally not required by the professional standard of care.

- On this project, if the Engineer had correctly identified the actual muck depths and quantities, bid price for the work might actually have been higher than the price negotiated in the change order. The large delay claim would probably have been avoided, but the final project costs would probably have been similar either way.

- Where significant subsurface uncertainties exist, it is important to recognize them and educate the Owner about their potential impacts. The Engineer and Owner need to work together in deciding how much spending on exploration is justified in comparison to the risks.

- It is important for the Owner to understand that some level of uncertainty always remains after the investigations are complete. No practical level of subsurface investigations can eliminate all potential surprises that might impact the construction cost and schedule.

- Some Owners are pleased to get bids for substantially less than the Engineer's estimate, but like anything else in life, extraordinarily low prices should usually be taken with a grain of salt.

- The Engineer should alert the Owner to the pitfalls that can be associated with overly optimistic low bids, including an increased likelihood for disputes, claims, and change orders.

- Low bids based on early completion schedules should not be accepted without a clear understanding and allocation of the inherent risks. In this case, the Owner probably did implicitly accept some of the risks of the tight schedule when he accepted the attractively priced low bid.

- If the Owner accepts a very low bid, he or she should retain a significant construction budget contingency for resolving the inevitable problems that will arise, and should be prepared for an increased amount of field inspection and resident engineering effort. On this project, almost two-thirds of the two-year engineering budget was consumed during the first season.

- In summary, the apparent up-front cost savings from a very low bid can be attractive, but due to the realities of construction, about which it is sometimes said that the only certainty is change, the final price is rarely as low as initially hoped.

Deep Excavation Practice in Fuzhou Soft Soil

By Zhang Yaonian[1] and Liu Xi-an[2]

Abstract: This paper presents a deep excavation case study of a project in Fuzhou City. Fuzhou is located on the southeast coast of China. The excavation covers a plan area of about 7000-sq. meters and extends to a depth of 10.7 m. The site is underlain by soft soils composed of recent fluvio-marine sediments. The design and construction of the retaining structures are discussed in the paper. It is shown that soldier piles with internal bracing system are appropriate retaining structures for deep excavations in Fuzhou.

INTRODUCTION

Due to the rapid economic growth and urban development in Fuzhou during this decade many new tall buildings have been constructed. These buildings are required to incorporate at least one level of basement. But generally, 2 or 3 levels of basements are common. Deep excavation therefore becomes a part of the construction process.

Fuzhou is located on the southeast coast of China. In most areas of this district the subgrade of a structure will be located in soft soils of recent fluvio-marine sediment. Muck and mucky clay are encountered at shallow depths. Because of difficult soil condition and lack of sufficient local experience, deep excavation is a very challenging part of the work in Fuzhou.

This paper describes a deep excavation completed in the urban center of Fuzhou City. The building incorporates three levels of basements, which cover a plan area of about 7000 sq. m. Deep excavation was required to a depth of 10.7 m. and involved

[1] Fujian Institute of Building Research, Fuzhou, P.R. of China
[2] Fujian Institute of Building Research, Fuzhou, P.R. of China

removing material from a soft muck formation having a total thickness of more than 30 m. Soldier piles having two levels of internal bracing supported the excavation. Due to an owner specified requirement the excavation was carried out in two stages. Details of the excavation, e.g., site investigation, design of the retaining structure, construction processes and the results of monitoring, are presented in the following sections.

GEOTECHNICAL CONDITION

Detailed geotechnical investigation, including exploratory borings, static cone penetration tests, standard penetration tests and routine soil property tests, were conducted at the site. Site investigations revealed a soil profile as shown in Fig. 1. The subsoil condition of the upper portion of the site is relatively uniform. It can be described as consisting of a thin layer, about 2 m thick, of miscellaneous fill. Beneath the fill is a flow plastic muck to a depth of 18 m. The strata underlying the muck layer are interbedded fine sands and clay-muck to a depth of 30 m. The soil properties of these two major formations, Table 1, govern the excavation situation.

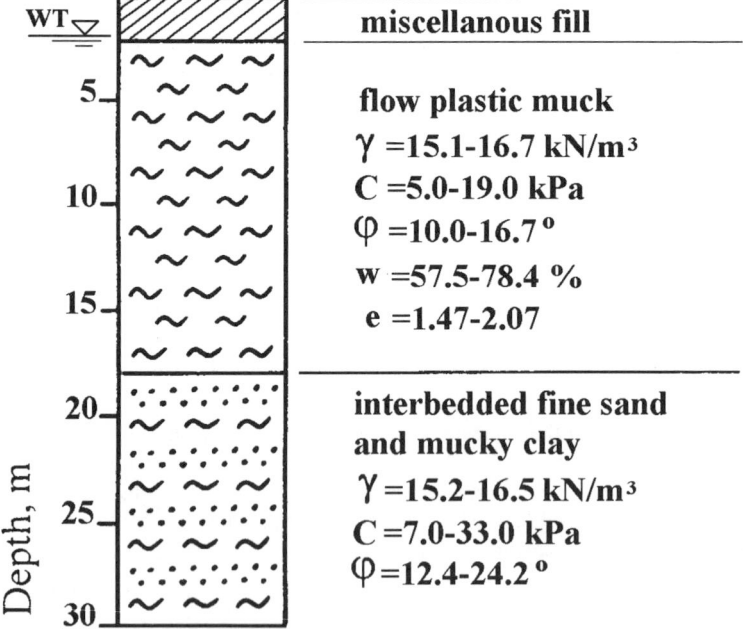

FIG. 1. Soil profile of the site.

Table 1 Soil properties in the site

Soil properties	Muck	Interbedded Fine sand and mucky clay
natural water content, %	57.5-78.4	59.5-70.5
natural unit weight, kN/m^3	15.1-16.7	15.2-16.5
natural void ratio	1.47-2.07	1.59-1.96
liquid limit, %	41.8-55.9	38.2-60.2
plastic limit, %	25.0-34.4	26.9-39.5
plasticity index	15.3-22.8	11.3-25.6
liquid index	1.08-2.91	1.09-3.68
modulus of deformation, MPa	1.60-3.08	2.13-3.09
cohesion of soil, kPa	5.0-19.0	7.0-33.0
angle of internal friction, °	10.0-16.7	12.4-24.2
SPT counts	1.3-2.4	2.4-2.8
coefficient of permeability, cm/s	9.12×10^{-8} - 2.80×10^{-7}	

RETAINING STRUCTURE DESIGN

Soil pressure

Soil pressure on the retaining structure was calculated according to classical Rankin theory. The selected soil properties and the soil pressure surrounding the retaining structure are shown in Fig. 2.

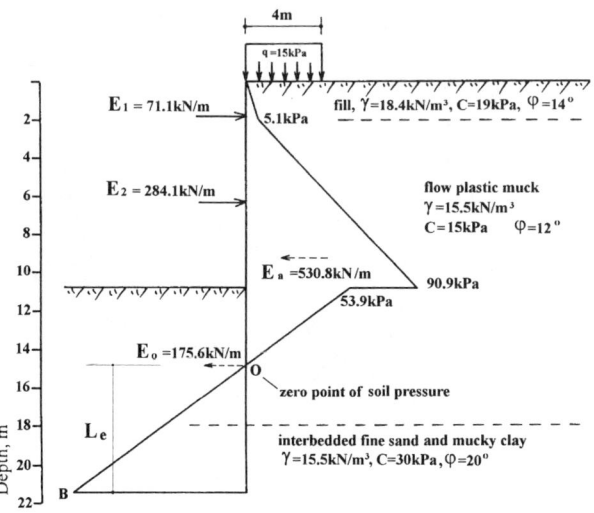

FIG. 2. Analytic model for excavation.

Mechanical analysis of excavation

As shown in Fig. 2, the zero point of soil pressure is considered as the support point of moment equilibrium. The reaction of the first strut was calculated as $E_1=71.1$kN/m when the excavated depth is 6.8 m. The reaction of the second strut was calculated in turn as $E_2=284.1$kN/m when the excavated depth is 10.7 m. The shear on the retaining structure at the position of zero soil pressure point is $E_0=E_a-E_1-E_2=175.6$kN/m. Considering the equilibrium of the pile segment below point O, we find that the summation of the force gives $E_0-BL_eL_e/2=0$. Hence the embedded length of retaining structure is $L_e=(2E_0/B)^{1/2}$. Here B is slope of soil pressure curve OB. The maximum moment in retaining structure is 584.6kN.m/m. It happens at the depth of 9.92 m.

Retaining structure system

The designed system consisted of retaining soldier piles with two levels of internal bracing struts. The soldier piles were 800 mm in diameter, 20.5 m in length and arranged at intervals of 300 mm. Since the coefficient of permeability of the excavated soil is less than 2.8×10^{-7} cm/s, waterproofing of the retaining structure was considered unnecessary. Because the excavation was of irregular quadrilateral and it was a requirement that it be excavated in two parts, the internal bracing system was designed as shown in Fig. 3. The elevations of two levels of internal struts are shown in Fig. 4.

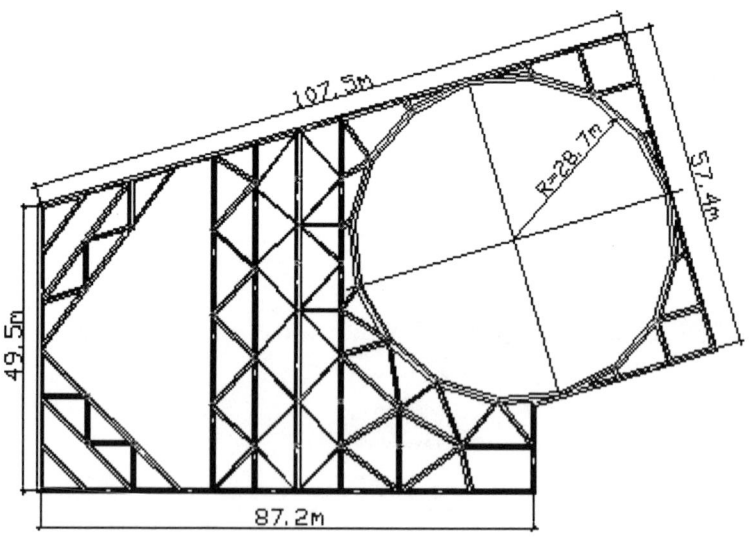

FIG. 3. Plan of internal bracing struts.

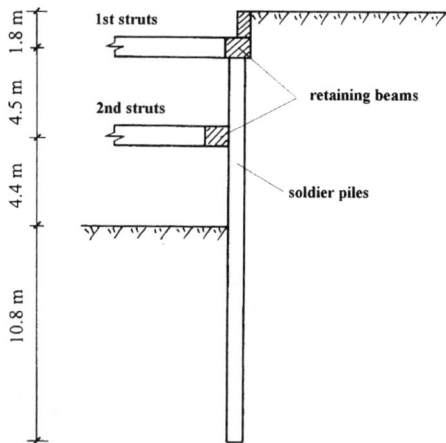

FIG. 4. Elevation of internal bracing struts.

CONSTRUCTION PROCESSES

The soldier piles were cast-in-place bored piles constructed using the slurry displacement method. The internal bracing struts were cast-in-place reinforcement concrete beams supported on steel columns. The construction processes of excavating and retaining were as follows:

(a) excavation was carried out to a depth of 2.3 m. and the first internal bracing struts were installed

(b) excavation continued to a depth of 6.8 m. and the second internal bracing struts were installed

(c) building foundation was constructed

(d) second strut was removed.

The construction processes of excavation are illustrated in fig. 5.

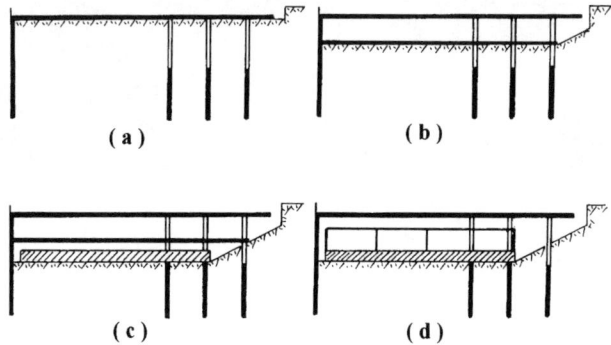

FIG. 5. Excavation sequence.

Because of the investment arrangements by the owner, it was required that the excavation be conducted by two stages. Fig. 6 and Fig. 7 show the completion of two excavations respectively. The west part of the excavation was conducted first. When that excavation was conducted, there was no retaining measure at the boundary of two parts. The surface of east part formed a relative stable slope. After completing the west part of the excavation, a temporary retaining wall was constructed, in case of an earth collapse, along the boundary between the two parts.

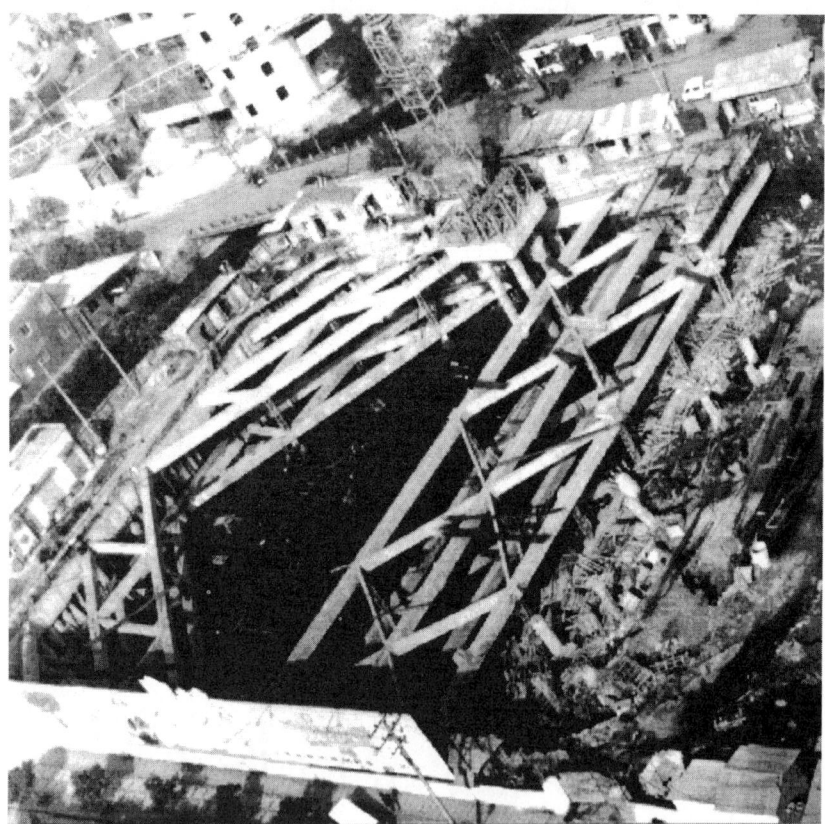

FIG. 6. Completed west part of the excavation.

EXCAVATION MONITORING

In order to guarantee the safety of the excavation, the construction processes were conducted under the guidance of site monitoring. The most important monitoring was lateral displacement of retaining structure. In this site, sixteen inclinometer pipes were arranged around the periphery of the excavation. The pipes were installed immediately on reinforcement cages of retaining piles. The lateral movements were measured by means of a SINCO inclinometer. Measurements were made regularly during the period the excavation was open. The measurement results indicate that the retaining piles moved laterally inward, i.e., towards the excavation, during the excavation in spite of the existence of the internal bracing reaction.

FIG. 7. Completed east part of the excavation.

When the final step of the excavation was completed, the heads of most retaining piles had moved inward about 20 mm. The maximum lateral displacement was about 40 mm in west part and 30 mm in east part; this maximum occurred at a depth of about 8 m below the first struts. The most significant lateral movement happened at the convex corner in the southeast part of the excavation. The maximum lateral displacement here was more than 70 mm during excavation. This location is the most unfavorable position for constructing and bracing the retaining structure. Fortunately, no significant damage occurred to the retaining structure from these lateral displacements.

SUMMARY

The excavation was completed without any effect on nearby buildings and municipal facilities. From the practice we have the following understanding.

1. Soldier piles with an internal bracing system are an appropriate retaining structure for deep excavation in Fuzhou. This system has the advantage of low cost, simple construction and safety.

2. Compared with traditional cross bracing system, the circular shape bracing system adopted in the excavation has many advantages. It provides a large space for construction of the excavation and the building basement. This project demonstrated that the system is reliable although the design was solely empirical.

3. The 300 mm interval between adjacent retaining piles is allowable. Seepage and soil collapse did not happen at the pile interval during the excavation. After the excavation was completed, the east part was exposed to the weather without any protective measures for more than one year.

4. Lateral displacement monitoring of retaining structure is the most important measure to guarantee the excavation will be conducted safely, because it provides immediate data on retaining structure behavior. Large displacements mean high strains in the structure. High strain means high risk. The builder must be prepared to adjust the excavation according to monitoring results.

INTAKE TOWER SEISMIC UPGRADE WITH POST-TENSIONED ANCHORS:
DESIGN AND CONSTRUCTION ASPECTS

Theodore B. Feldsher, P.E.[1]
Paul Krumm[2]
Greg Rollins, P.E.[3]
Laurel Harrington, P.E.[4]

Abstract

The South Fork Tolt Dam Intake Tower was designed and built in accordance with the standards of practice current in the early 1960's. Thirty years later, analysis based on modern structural design criteria and updated seismic hazard data concluded that the tower was highly vulnerable to earthquake damage and needed strengthening. Due to the critical importance of the water supply, the Owner mandated that the reservoir remain in service and that water quality be protected during the repair work. The innovative seismic upgrade design called for six high-capacity post-tensioned anchors, installed into holes drilled with high accuracy down through the thin walls of the tall, slender structure. Specialized construction techniques and equipment were employed to successfully complete this challenging project, which required some of the largest and longest post-tensioned anchors ever used to seismically strengthen an intake tower structure.

INTRODUCTION AND BACKGROUND

Project Description

The South Fork Tolt River Project is located on the western slope of the Cascade Mountains in King County, Washington, about 30 miles inland from Seattle. The project is owned by Seattle Public Utilities (SPU), and is the water source for about 400,000 people in the Seattle area. The 60,000 acre-foot Tolt reservoir, retained by a 200-foot high earth dam, also provides about 17 megawatts of hydroelectric power. The reservoir level is controlled with a freestanding intake tower and a morning glory spillway tower, both of which are partially embedded in the upstream shell of the dam, as shown in Figure 1. The reinforced concrete intake tower was designed in

---

[1] Senior Engineer, URS Greiner Woodward Clyde, Oakland, CA
[2] Layne Christensen Company – Geotechnical Construction Division, Salt Lake City, UT
[3] Senior Engineer, URS Greiner Woodward Clyde, Portland, OR
[4] Supervising Civil Engineer, Seattle Public Utilities, Seattle, WA

1958; Figure 2 shows the tower near the end of construction in 1962. The tower has a total height of 179 feet above the foundation, with a top platform at Elev. 1776. The upper 120 feet of the tower has a narrow rectangular cross-section with gated intake openings at three different levels. The embedded portion of the tower, below El. 1655, contains the water supply conduit, low-level outlet, and control valve.

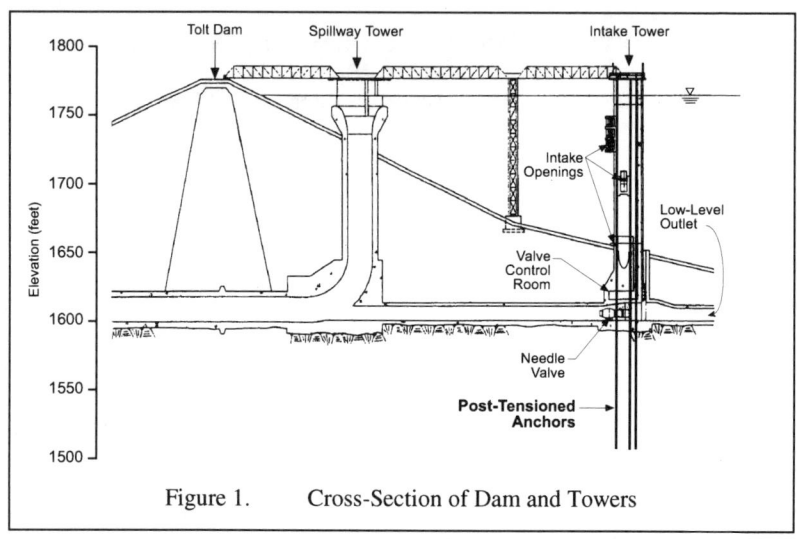

Figure 1.   Cross-Section of Dam and Towers

Seismic Safety Evaluation

In 1992, SPU retained Woodward-Clyde to carry out a safety evaluation of the dam and intake tower, and to develop earthquake ground motion parameters for the analyses. The results showed that the dam was sound, but indicated that the intake tower had significant structural weakness and a high potential for total collapse during a large earthquake. The major source of weakness was identified as inadequately lap-spliced vertical reinforcing bars in the tower walls. The short lap splices were present at construction joints near Elev. 1655, where #14 and #18 bars joined. This elevation is near the base of the freestanding portion of the tower, which is the critical section for earthquake-induced bending moments. The safety evaluation concluded that heavy damage or even collapse of the tower could occur due to brittle failure in tension at the inadequate splices. Such a collapse would disable the only means for lowering the reservoir level, and would pose an unacceptable dam safety hazard.

Performance Objectives and Design Earthquakes

Based on the unacceptable consequences of tower failure, SPU decided to proceed with a seismic upgrade, to ensure their ability to safely drain the reservoir if needed after the Safety Evaluation Earthquake (SEE). SPU's secondary objective for the upgrade is to maintain the reservoir in full, unimpaired operation after an Operating Basis Earthquake (OBE). SPU also set an intermediate objective for maintaining operability with limited damage after a 500-year Earthquake, because of the facility's lifeline designation. For design purposes, the SEE, 500-year earthquake, and OBE were determined to generate peak horizontal ground accelerations at the site of 0.50g, 0.34g, and 0.20g, respectively. The SEE was defined based on a magnitude 6.5 earthquake centered 10 km from the site.

Figure 2. Tower Construction, 1962

SEISMIC UPGRADE DESIGN

Design Criteria and Main Elements

Design alternatives for strengthening the intake tower were evaluated on the basis of risk and effectiveness, technical feasibility, design precedent, constructability, and cost. Constructability considerations were paramount, because SPU required the reservoir and intake to remain in operation during construction, and stringent water quality requirements allowed for no disturbance within the reservoir itself. The final design includes six 58-strand post-tensioned anchors, each with a design load of 2060 kips. These anchors act to stiffen, strengthen, and tie down the tower structure. They provide a 580 psi axial prestress load and act as supplemental tensile reinforcing across the critical construction joints. As shown in Figure 1, each anchor is installed in a 275-foot long, 11-inch diameter hole, drilled down through the tower walls and into the foundation bedrock. The design also includes a new watertight outlet valve operator and power supply modifications, so the low-level outlet will remain operational if the control room at the base of the tower floods due to cracking above.

Anchor Layout and Drilling Alignment Tolerances

The final anchor layout was developed to keep the centroid of the anchor forces as close as possible to the centroid of the structure, as shown in Figure 3. The anchor layout options were limited by the 24-inch thick tower walls, and by the presence of the stairwell, intake shaft, intake gates, vent pipes, and the low-level outlet gate and needle valve at the base of the tower. To maintain minimum clearances, a very tight 1:200 alignment tolerance was specified. Hole curvature tolerances were also specified, to maintain a 0.5-inch minimum cover between the outermost anchor strands and the borehole wall after stressing. These alignment tolerances allow each anchor hole to deviate no more than 11 inches from vertical at the base of the 180-foot high tower. Because of the potentially serious consequences of drilling alignment errors, 3-inch pilot holes were required at each anchor location, and frequent downhole alignment surveys were required during pilot hole drilling.

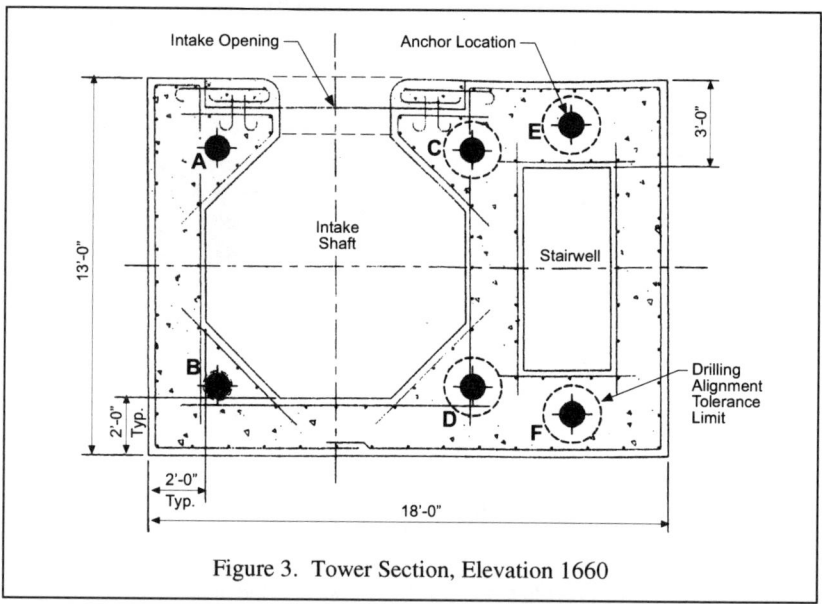

Figure 3. Tower Section, Elevation 1660

Anchor Strands and Corrosion Protection

Each anchor contains 58 strands of 0.6-inch diameter, 7-wire, low-relaxation Grade 270 steel. Epoxy coating and epoxy filling in accordance with ASTM A-882 was specified as the primary corrosion protection measure for the anchor strands. Other corrosion protection measures such as corrugated plastic sheathing were judged unfeasible given the length and diameter of the anchors. Epoxy-protected strand was also selected based on reported experiences from other large multistrand anchor projects. To compensate for the greater relaxation load losses associated with epoxy-protected strand, the design called for a 4-week load relaxation period following initial stressing and lockoff of each anchor. Each anchor is then checked for load loss and restressed if needed to compensate for any relaxation.

Bond Length Design

Load transfer from the strands through the grout into the rock mass surrounding the bond zone resists the tensile load in each anchor. The most critical interface in the bond zone is typically between the borehole wall and the grout. The actual bond stresses present along this interface are expected to be highly nonlinear and dependent on the shear strength of the rock and the grout, the jointing in the rock mass, the roughness of the borehole wall, and other factors. For design purposes, average stress values for the entire bond length are usually used, with some reduction in bond length to allow for debonding. For design of the intake tower anchors, bond stress values of 250 psi (ultimate) and 80 psi (allowable) were used. These are based on values reported in the literature for similar rock types, and on the results from two small-scale test anchors installed in 3-inch boreholes and tested to failure. Based on the established 80-psi allowable stress level, an 80-foot bond length was selected.

Fully Bonded Design

Unlike many anchoring projects, the intake tower anchors were designed to be fully bonded to both the structure and foundation. This was accomplished by secondary grouting the free length of each anchor after completion of final restressing. Once fully bonded, the stressing loads are locked into each anchor, and further relaxation losses are very limited. Although future restressing is not possible, the fully-bonded design was selected because it provides an increased ultimate strength compared to the unbonded alternative. The higher ultimate strength provides a greater degree of overall structural ductility, because the bonded anchors act to distribute rupture strains and cracking along the length of the structure, reducing the potential for premature concentrated cracking and localized crushing failures during seismic loading.

Structural Analyses and Expected Performance

Static and dynamic finite-element structural analyses were carried out to verify that the strengthened tower will meet the performance objectives. A simplified "stick" model of the tower was used for the analyses, which were performed assuming both linear elastic and nonlinear material properties for the structure. For the OBE, the dynamic time history analysis results indicated satisfactory behavior with little or no damage. For the 500-year earthquake, the linear elastic analysis indicated some minor cracking and spalling, but not enough to impact either the operation or overall safety of the structure. Some damage repair is likely to be needed, but not immediately. For the SEE, which was the largest earthquake studied, the linear elastic analysis showed bending moments significantly exceeding the ultimate capacity of the strengthened tower. However, this result was judged not to be representative of the expected actual behavior, since the tower properties change as damage, cracking, and steel yielding occur. The nonlinear analysis is more representative for this case, because it allows the model to account for changes in the properties. This analysis showed SEE-induced bending moments greater than the elastic capacity, but approaching the ultimate capacity for only a few cycles of shaking. On this basis, little inelastic deformation was predicted, and collapse is not expected. However, the strengthened tower is expected to undergo significant cracking and spalling of the concrete, particularly near El. 1655. The cracking may result in flooding of the stairwell, but this is not expected to impact the overall stability or post-earthquake safety of the structure. Overall, the analysis results showed that the reinforcing and anchoring design should strengthen the tower enough to survive the design earthquake and to meet the established seismic performance objectives.

PROJECT CONSTRUCTION

General

Because of the unusual and difficult nature of the project, SPU decided to prequalify specialty subcontractors for the drilling and anchor work in advance of bidding. Prime contractors were then required to select from among the prequalified subcontractors bid time. A total of four bids for construction were received in February 1997, ranging from $4.3 million to $5.3 million. For comparison, the Engineer's Estimate was $3.4 million. The Contract was awarded to the low bidder, who named Layne Christensen Company as their subcontractor for the drilling and anchor work. Notice to proceed was given in May 1997. The post-tensioned anchor work started in September 1997, and was finally completed in August 1998. The final construction cost for the tower seismic upgrade and related work, including change orders, was about $5 million. Numerous challenges had to be overcome during construction, including environmental protection, limited access, difficult drilling, winter weather, and handling of the very large anchors. The construction contract provided for a Disputes Review Board, along with Escrow of Bid Documents for use in dispute resolution. Generally, these measures provided motivation for timely resolution of disputes in the field. However, one substantial claim from the General Contractor remained unresolved at the time this paper was prepared.

Environmental Protection

The construction contract required installation of a special floating turbidity control curtain, as shown in Figure 4. This curtain was anchored and sealed to the tower walls about 30 feet below the surface, thereby completely encapsulating the water around the tower work area. This greatly reduced the risk of releasing any drilling fluid spill into the reservoir. Drilling fluids and cuttings were collected in special containment tanks on the tower deck, and pumped to settlement ponds downstream of the dam. Continuous automated water quality monitoring equipment was installed, with an audible alarm system. Through a combination of caution and vigilance, the reservoir remained in service during construction with no major water quality impacts.

Access and Logistics

The only existing access to the tower from the dam's crest is via a steel truss footbridge, 300 feet long by 5 feet wide, with a maximum point load rating of 1000 pounds. This required all heavy equipment and tooling to be ferried from a dock located one-quarter mile distant from the tower. The existing tower platform deck measures only 20 by 25 feet, and the available working area is congested by the presence of three hydraulically operated intake gate hoists, a reservoir outlet gate hoist, a hydraulic power unit, the intake shaft access hatch, and the stairwell

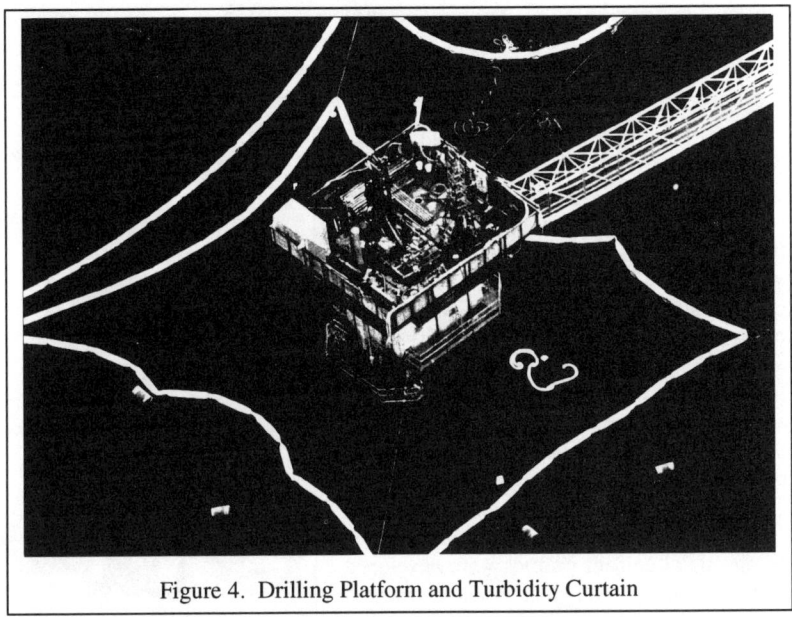

Figure 4. Drilling Platform and Turbidity Curtain

opening. To overcome this lack of working space, the contractor erected a temporary 30 by 30-foot structural steel platform 9 feet above the tower deck elevation. This platform, shown in Figure 4, provided an unobstructed area to stage the drills and support equipment. An 80-ton crane mounted on the transport barge was used to hoist equipment and supplies onto the drilling platform. Because the available work area was still quite limited, it was crucial to plan the layout of equipment prior to moving anything onto the deck. To optimize space on the platform, the contractor used scaled overlays and CAD drawing software to determine the most efficient layouts for his drills and support equipment.

Anchor Hole Drilling and Preparation

As shown in Figure 3, the seismic retrofit design called for four anchors around the intake shaft and two on opposite sides of the adjacent stairwell. The four anchor holes around the intake shaft penetrated the ceiling of the control room at 150-foot depth, recollared into the control room floor, and continued through the needle valve chamber and into the underlying bedrock. After each anchor hole was drilled to within the specified alignment tolerance, the bond zones were water pressure tested to detect fractures and joints in the rock. All but one of the holes required subsequent grouting for watertightness, followed by redrilling. The design locations for anchor holes A and B coincided with existing 6-inch and 12-inch vertical vent pipes embedded in the tower walls. The design called for abandoning those portions of the vents above the control room and incorporating them as part of the new anchor holes. This proved more troublesome to accomplish than originally envisioned, because 6-inch vent pipe was not as straight as expected, which led to problems associated with negotiating the bends with the reaming tools.

Pilot hole drilling was conducted using a diamond-impregnated drill bit with an HX-sized (3") wireline double-tube core barrel system. The pilot holes were advanced for the full depth of each anchor (except at the vent pipes, where pilot holes were unnecessary), ranging from 275 to 280 feet below the top of the intake tower. The hole alignments were closely monitored during pilot drilling with high-precision gyroscopic surveying equipment. As the holes advanced through the tower walls, gyroscopic surveys provided accurate orientation measurements of the hole angle and direction (azimuth) at 10-ft. depth intervals. From the surveys, composite alignment profiles were developed using a laptop computer and specialized directional drilling software. The survey trajectory data was integrated with CAD software, and the drilled holes were plotted

Figure 5. Stepped Reaming Bit

with respect to the tower structure. The pilot holes were drilled with specially manufactured fluted core barrels, diamond impregnated core bits, stabilized drill strings, and controlled drilling parameters (including rod rotation and weight-on-bit). The drilling accuracy achieved was such that when the first pilot hole penetrated the ceiling of the control room (150 feet below the platform), it was within an inch from the target center.

Due to concerns about potential impacts to the thin-walled tower concrete, the specifications prohibited use of percussion drilling methods within the tower structure. Rotary drilling methods were used instead, with successive passes of specially built reaming bits. The reaming bits were made from stacked diamond-tipped coring bits, fitted with a nose guide to follow the pilot hole, as shown in

Figure 5. Reaming was performed in three stages: from 3 inches to 6-7/8 inches, then to 9-1/8 inches, and then to the final diameter of 11-3/8 inches. The stepped reaming bits were manufactured with the same approximate kerf area for each successive reaming pass, to maintain consistency in penetration rates. Once the full-sized holes reached bedrock, the drilling equipment was switched to a downhole percussion hammer. This provided for much faster drilling and resulted in rougher borehole walls with improved bonding capacity.

The 6-inch vertical vent pipe embedded in the tower wall at the design location of anchor hole A presented significant difficulties during drilling. Although it was known in advance that the vent pipe was not precisely straight, it was not known in advance that the drill rods would be too stiff to readily negotiate the bends. This significantly impacted overall drilling progress, as removal of stuck drill rods was required, followed by backfill grouting and redrilling a straighter pilot hole. Once the pilot hole was in place, the existing embedded steel vent pipe presented further difficulties, as it had to be overcored and removed in short sections. The 12-inch vent pipe at hole B was much easier to deal with, since it did not have to be removed, and required only a thorough cleaning prior to normal anchor hole drilling through the rest of the tower below the control room.

The three electric-powered hydraulic drills utilized on the project included a Diamec 260 for the pilot holes, and for the full-sized hole reaming, a modified Terramec 1000 and a Longyear 38. The pilot hole drilling required 4 months to complete, and nearly 7 months were required for the reaming. The longer time period required for the reaming was due to the additional passes (3 in all) required to achieve the final anchor hole diameter. Drilling progress rates ranged as high as 40 to 50 feet per shift for the pilot holes, but usually did not exceed 20 to 25 feet per shift for each reaming pass on the full-sized holes.

Figure 6. Anchor Fabrication

Anchor Fabrication

As fabricated, each anchor was nearly 300 feet long, weighed close to 15,000 pounds, and contained almost 17,400 feet of 0.6-inch diameter cable. Combined, nearly 20 miles of cable were used to make the six anchors. Because of the difficulties associated with coiling and transporting such long and heavy anchor

Figure 7. 58-Strand Anchor Head

bundles, the supplier opted to fabricate the anchors on-site. This decision contributed greatly to minimizing handling damage to the epoxy coating on the strands. The epoxy-coated, grit-impregnated, high-strength steel anchor strand was delivered to the site on large spools. The strands were unreeled through a series of specially-fabricated HDPE organizers and centralizers, as shown in Figure 6, to ensure minimum clearance requirements for grout cover. Three grout tubes were included in each anchor bundle for primary and secondary grouting. The bottom ends of the strands on each anchor were grouted into protective steel nose cones to facilitate smooth insertion into the boreholes. The anchor head hardware included 8-inch thick steel wedge plates and the permanent gripping wedges. Each wedge plate was fabricated to accommodate the 58 strands and 3 grout tubes, as shown in Figure 7. For hoisting, the anchor heads were equipped with special wedge retainer plates and eyebolts so lifting shackles could be attached. The anchors were fabricated and assembled by a five-man crew in a total of about eight weeks.

Anchor Installation

The originally proposed anchor installation procedure entailed using an off-coiler reel. This would have required individually coiling each anchor, ferrying the coil to the tower, and hoisting it onto an off-coiler unit on the drilling platform for installation. This process would have taken at least several weeks to complete for all six anchors. The contractor submitted a value engineering proposal requesting use of a helicopter for installation in lieu of the off-coiler. The helicopter offered the main advantage of greatly reduced installation time, and also had potential for increasing safety and minimizing handling damage. Helicopter use had originally been disallowed by SPU due to environmental concerns, but these were satisfactorily allayed and the proposal was subsequently accepted.

The contractor mobilized a Sikorsky S-64 Sky Crane helicopter for the anchor installation work, as shown in Figure 8. The helicopter was powered by two 4,500-shp jet turbines, was equipped with an anti-rotation cargo handling system, and was specifically designed to perform precision external load handling operations like installation of transmission line towers. According to the pilot, when hovering 350 feet in the air, the anchor holes looked about the size of a dime from two stories up. To help guide the anchors into the holes, a special two-part funnel was devised so that once the anchor started into the hole, the funnel halves could be removed allowing it to continue down the hole without interference. To minimize the potential for abrasion damage to the epoxy coating, the funnel was lined with a heavy carpet material.

Figure 8. Helicopter Installation

Special rigging techniques were eventually implemented to aid crews in controlling the motion of the anchors, after erratic lateral swinging of more than 30 feet thwarted the first several installation attempts. It took nearly four hours and many attempts to install the first anchor, two hours to install the second, and, on average, ten minutes for each of the remaining four. Once a procedure was developed, the fastest installation was accomplished in five minutes from pick to set, a testament to the proficiency of the ground and air crews. The decision to use the Sky Crane helicopter resulted in all the anchors being installed in a single day, cutting at least two weeks from the project schedule. Although helicopters have been used on other anchor installation projects, this may have been the first time for anchors of this size and length.

Anchor Stressing and Performance Testing

Each anchor was pre-equipped with separate tremie grout tubes for the primary and secondary grout zones, along with a backup tube. Following installation, each hole was thoroughly flushed for a final cleaning, and a specially formulated cement-based anchor grout was pumped through the grout tubes into the 80-foot bond zone of each anchor. The specifications required a 14-day minimum curing time and a 5,000-psi minimum compressive strength for the grout before stressing. Once the grout achieved the specified compressive strength, the anchors were tensioned in a staggered pattern around the tower, to meet the specified design load of 2060 kips. Lockoff and performance test loads were 2215 kips and 2530 kips, respectively.

The specifications originally required the contractor to stress the anchors using a stressing jack with sufficient stroke and capacity to accommodate the full expected elongations of the anchors under load. However, during construction the contractor submitted a value engineering proposal to use two jacks, placed one on top of the other, to achieve the very long stroke capacity required. The contractor reported that only a few stressing jacks large enough to meet the specifications existed in the world, and that significant delays in obtaining such a jack were possible since transport from overseas would likely be necessary. The proposal was subsequently accepted, and stressing proceeded with two stacked 1400-ton jacks, each with a 14-inch stroke length. The final jacking arrangement used is shown in Figure 9.

Figure 9. Anchor Stressing with Two Jacks

As a part of stressing, full performance tests were conducted for each anchor. Stressing was performed by extending one jack until the available stroke was nearly exhausted, and then extending the second jack until the target load was reached. Loads were recorded based on the pressure in the lower jack only. For the theoretical free stressing length (about 200 feet), the maximum elongations were calculated to be

17.25 inches. Actual recorded elongations at maximum load ranged from about 16 to 21 inches. Overall, the elongation data obtained during performance testing indicated that the anchors behaved elastically during stressing, with only small amounts of residual elongation and no creep. The six anchors were stressed over a period of 1½ weeks. This included one day to set up the stressing equipment and apply the alignment load, and one day to stress and lock-off each anchor. The anchors were locked-off at or above the specified load and then monitored with calibrated vibrating wire load cells.

During the design phase of the project, the number of strands in each anchor was increased from 54 to 58, without increasing the total anchor design load. This was done as a measure of conservatism, to provide a safety margin in case of any strand breakage, damage, or slippage during handling and installation. This safety margin was particularly important because of the extreme difficulty involved in removing any problematic anchor after installation, and because the limited tower area offered no room for installation of any replacement anchors. In the end, the conservative design paid off, because during or immediately following lock-off, three of the anchors each lost load in several strands. The mechanism for the load loss was slippage of the individual strands back through their 3-part gripping wedges. Several of the failed wedges were later removed and examined to see if the reason for the slippage could be ascertained. The teeth of the gripping wedges appeared to be sheared off in some areas and clogged with epoxy in other areas. Something had clearly prevented them from gaining a sufficient bite on the steel strand beneath. Some possibilities include an excessive epoxy coating thickness, inadequate gripping teeth design, some slight misalignment or twisting of the entire anchor bundle, or perhaps a combination of factors. Ultimately, most of the slipped strands were regripped with new wedges and restressed using a monostrand jack. The few that could not be restressed due to the inaccessibility of their wedges were deemed tolerable, because the extra four strands in each anchor allowed the total load to be carried without overstressing the remaining strands.

Since the total load applied to the tower by the tensioned anchors is roughly six times the tower's self weight, extensive monitoring was carried out to observe the tower behavior during and after anchor stressing. The instrumentation installed and monitored included 16 strain gauges, 4 extensometers, 2 tilt meters, 6 load cells, and 2 reflective survey monuments. These were placed on the top platform, within the stairwell, within the intake shaft, and on the outer surfaces of the tower. All instruments were monitored prior to, during, and after performance testing and stressing each anchor. Instrumentation data obtained during stressing indicated that tower deflections and strains were generally well within the acceptable range. The maximum lateral tower deflections observed during stressing were about 0.7 inches, only half of what had been predicted by finite element analyses performed during design.

Following the specified 30-day evaluation and load relaxation period, only three of the six anchors required restressing and addition of shims to bring them back into the specified lockoff load range. After restressing, the free lengths of all the anchors were fully grouted to permanently bond the anchors to the structure, and the anchorage hardware was permanently capped with weatherproof covers.

SUMMARY

The disciplines of both seismic hazard assessment and structural analysis have made substantial advances since the late 1950's, when the Tolt Intake Tower was designed. Safety evaluations carried out to present-day standards showed the tower to be highly vulnerable to damage and possible collapse during a major earthquake. The seismic strengthening design that was subsequently developed to remediate the problem called for very long, very high-capacity post-tensioned anchors, installed in holes drilled to very tight alignment tolerances. The specialty subcontractor used both innovative approaches and tested techniques to surmount these and other difficulties. In the process, he achieved precedent-setting results and maintained full compliance with very stringent environmental requirements.   The successfully completed project represents what may be the first-ever use of such large and long post-tensioned anchors to strengthen a tall, thin, and seismically vulnerable intake tower structure.

Subject Index

Page number refers to the first page of paper

Author Index

Page number refers to the first page of paper